Are We There Yet? The Search for a Theory of Everything

EDITOR

Moataz H. Emam
State University of New York College at Cortland
USA

eBooks End User License Agreement

CONTENTS

FOREWORD

Every epoch has its great scientific questions. It is our hubris at the moment to dream that we can figure out the final theory of the fundamental interactions of physics. This dream goes back to Einstein and his search for the *Unified Field Theory*, for which he devoted the latter part of his life. However, in retrospect it seems clear that Einstein lived too early for this task. At that time we as a field did not yet even know how to describe the known interactions, or how to use quantum field theory properly. Einstein did not have the right knowledge base and tools to complete his task. These would only come later. Presently we are clearly in a better position because we have a rich and successful description of the world encoded in the Standard Model. We have learned much about Nature and are ready to renew Einstein's dream for a final theory. Is it our time?

This book is a collection of reflections by young scientists on the goal of a final theory. The science is discussed without its mathematical underpinnings, so that the essays are suitable for a general reader. They are similar to polished versions of the discussions that one hears around the lunch table, debating the dreams and prospects of the unexplored pathways ahead. The issues are subtle and interesting.

The descriptions of the known interactions themselves give us a strong clue towards unification. They point to a narrow range of energies in which the unification would occur. This is most obvious for the theory of gravity, which is why so much attention is paid to that interaction. Contrary to old-fashioned wisdom, gravity and quantum mechanics are not incompatible, as has been said for many years. Recent understanding in quantum field theory, in particular the development of *effective field theory*, shows that gravity and quantum mechanics work perfectly naturally together at low energies. Quantum calculations can be made that predict very tiny modifications of the classical predictions of general relativity. However, these corrections grow with the energy, become overwhelmingly large at an energy referred to as the *Planck energy* (or Planck mass or simply Planck scale). The well-behaved theory appears to be falling apart above this energy. Something is clearly happening at that scale – probably a new description is needed there. Moreover, the other interactions also point to an energy scale close to the Planck energy. The strength of their interactions also depend on energy, and the electromagnetic, weak and strong interaction strengths all become approximately equal just below the Planck scale. Therefore that scale is where we would expect unification to occur. So we think we have a clue as to where the action is.

However, that unification scale is not readily accessible. The Planck energy is about 10^{16} times *higher* than the energy of the Large Hadron Collider (LHC), the most energetic device in science. We will never be able to run new experiments at the energy of unification. This is the great hurdle to our present dream. So maybe our problem is different from Einstein's. Even if you grant that we may possibly have the theoretical tools to describe unification, will we have the experimental ability to learn about Nature at the unification scale? It is actually not hopeless despite the large gap in energy. The early universe may reveal effects from the unification era. Studies of very rare processes, such as the possible decay of protons, may be sensitive to effects from very high energies. However, it has to be admitted that we too may fail at our dream if we are not clever enough to figure out how to validate our theories of unification.

This disparity between the available theoretical tools and the intrinsic obstacles to experimental verification underscores some aspects of the sociology of the field at the moment. There is a sometimes bitter debate over the directions of the field and this debate is played out in the public literature and blogs. It is more gently visible within the essays of this book also. That the debate can become so heated is at least partially because experiment is not riding to the rescue. In situations where experiment will quickly decide which of competing theories is correct, people do not need to be overly attached to their theory – it could well be proven wrong next week. But when one does not know if one's theoretical work will be tested within one's career, the need to feel correct in one's research direction (and thereby feeling that others are incorrect in theirs) becomes higher (This comment is not my own – I heard it ascribed to James "BJ" Bjorken. I repeat it because I feel that it explains some of the nature of the public debate). However, it should be said that the fact that the theories are so hard to test is not the fault of the theorists themselves. Many try mightily to think of feasible ways of testing their ideas. The difficulty is just the intrinsic nature of unification theories, required by the remoteness of the unification scale.

One variant of the dream would actually be readily verified. This would be if the unified theory predicted all the parameters of the physical world without any adjustable assumptions. There are about 28 numbers describing the Standard Model plus gravity – these give the masses of the particles, the strength of the interaction and various properties in the decay of unstable particles. They appear in the theory without any obvious order – for example the mass of the top quark is about 45,000 times the mass of the up quark, with others falling seemingly randomly (technically, random on a logarithmic scale) in between. No one has made much sense of the pattern of these parameters. They are not predicted by the Standard Model itself – presumably they are the input from some more fundamental theory. Indeed, they appear to be non-trivially odd, in that small variations in the parameters would lead to worlds without atoms or nuclei or without stars. The fact that the parameters have only a small *anthropic* range – i.e. those compatible with life as we know it – is an interesting topic in its own right and has a modest scientific literature. In any case, we have all these numbers which we don't understand but which appear to be *just so* for the existence of key features of our universe. It would be lovely to explain them.

The longtime dream of String Theory, which has essentially no free parameters, was that once we could solve the theory we would predict exactly all of these parameters. If that were to work it would be outstandingly impressive and would readily be taken as a verification of the theory. This particular dream has faltered from the indications that String Theory may have so many solutions (the so-called *landscape*) that *any* combination of parameters should be possible, and indeed there would be many other realizations besides the Standard Model. At one level, this is deeply discouraging – there goes our big test. However, if that is the nature of the theory, we may have to live with it. This changes the way that we approach the theory of the early universe. Are there domains in the universe that have different values of these parameters, and if so does this explain the special values that we see in our world of atoms and stars? It also changes the way we look for tests of the theory. Does string theory predict that the all the masses should be random on a logarithmic scale or not? Maybe some of the other theories will come to the rescue here and uniquely explain all the features that we observe in the Standard Model.

It is of course possible that none of the possibilities under consideration are viable. We could be in the same situation as Einstein was, lacking the knowledge and tools to even head in the right direction. Most directions at present are based on the idea of unification of symmetries. Our present theories have certain symmetries, and the postulate is that the fundamental theory would have more symmetry at high energy. But perhaps the reverse is true – maybe the symmetries that we see are artifacts at low energy, emergent from a fundamental theory without them. This would upset almost all work being done presently. It is important to keep an open mind to all possibilities.

We could have some hints in the next few years, coming from the working of the LHC. Although the LHC is unable to directly probe unification at the Planck scale, it may give us hints as to the correct direction. For example *Supersymmetry* is a key ingredient to String Theory – if evidence for this is found at the LHC it would be an enormous boost for the direction of String Theory even though it does not directly confirm the theory. If it is not found, it is not fatal for the theory but certainly would be discouraging. Moreover, all past colliders that opened up new territory have also found unexpected developments. If that happens, we could get enormously valuable clues. The most optimistic outcome would be if we could find evidence that the unification scale is lower in energy than the Planck scale. So we have the pleasure of looking forward to exciting developments in the next few years.

Voltaire's philosophical satire "Candide, or Optimism" tells the story of the eternally optimistic but misguided professor of "metaphysico-theologo-cosmolonigology", Pangloss, who argues that we live in the best of all possible worlds, a belief based on Leibniz' theoretical arguments. With Candide, Pangloss makes an excursion beyond his castle, finding evidence otherwise in some of the atrocities encountered. The book ends with the enigmatic words "We must cultivate our garden" (thereby spawning thousands of college essays to explicate their meaning). I have always had a fondness for this fable, especially when discussing anthropic issues. However, in the context of the present volume, the phrase "cultivate our garden" has a special meaning. We really don't know which of our present directions will be fruitful. In research, many seeds are planted but only a few really blossom. Each of us tries to cultivate our own corner of the garden in order to see what grows. The young scientists whose essays you are about to read provide us with a tour of their gardens, revealing their thoughts about the prospects and pitfalls for the future.

John F. Donoghue
Amherst, MA

PREFACE

When I was asked by Bentham Science Publishers to propose an idea for an e-book on current issues in physics I thought it was a no-brainer. For several decades the ambition of unifying all of physics under one fundamental set of rules that "fit on one side of a T-shirt" (as some physicists are fond of saying) has been an increasingly nagging objective of basic research. Many ideas were proposed, and this book explores some of the most recent of them, but there has never been a more exciting time in the history of this search: Our most complex experimental tool yet; the $10 billion "Large Hadron Collider", in Geneva Switzerland, has been turned on and has already had some results published. Who knows what wonderful things it will produce over the next few years and in what directions of research will it guide us. It is then the perfect time to slow down, take a deep breath, and ask ourselves "Are we there yet?" This is what this book is about.

The book is (roughly) structured following the theme:

History ⇨ *Articles on String Theory* ⇨ *Articles on other theories (Loop Quantum Gravity etc.)* ⇨ *New ideas and speculations* ⇨ *General discussions*

The articles, however, are self-contained, hence the readers may choose to read them in any order they feel comfortable with. Inevitably, there is some overlap in the material as different authors briefly review the same concepts. This is done, however, using different styles and approaches and is not repetitious. In most of these articles, the authors present their own views on what the future will be like and what possible new tools physicists may need. The following are brief biographical notes about the contributors, in order of appearance:

- *The Foreword* was written by John Donoghue; professor of theoretical physics at the University of Massachusetts at Amherst. His expertise ranges over several branches of physics at once, having published papers in six research fields: theoretical high energy physics, phenomenology of high energy physics, general relativity, quantum gravity, astrophysics and nuclear physics. He is currently developing ideas in the field of emergent physics. Professor Donoghue is co-author to a classic textbook on quantum field theory "Dynamics of the Standard Model", published by Cambridge Monographs on Particle Physics, Nuclear Physics and Cosmology" (1992) and editor of "CP Violation and the Limits of the Standard Model (TASI 1994)" published by "World Scientific" (1995). (*http://blogs.umass.edu/donoghue/*).

- "*The Fire in the Equations*". Tasneem Zehra Husain is a string theorist; currently assistant professor at the LUMS School of Science & Engineering, Sector U, D.H.A, Lahore, Pakistan. She received her PhD from Stockholm University, followed by a postdoctoral research position for two years at Harvard University. Prior to that she had acquired a scholarship at the International Center for Theoretical Physics (ICTP) in Trieste, Italy. Her research interests have so far focused on eleven dimensional supergravity theory. Dr. Husain is the first Pakistani woman to ever achieve such high accomplishments in science.

- "*String Theory and the Failure of Unification*". Peter Woit is a mathematical physicist, currently Senior Lecturer in the mathematics department at Columbia University. After a doctoral degree in particle theory from Princeton, he was a postdoctoral researcher at the YITP in Stony Brook and MSRI in Berkeley before arriving at Columbia. His book "Not Even Wrong" was published in 2006, and he has been maintaining a blog by the same name since 2004 at (*http://www.math.columbia.edu/~woit/blog*).

- "*Can String Theory Survive Complete Falsification?*" Moataz H. Emam is assistant professor of physics at the State University of New York College at Cortland, where he focuses his research on particular aspects of symmetries that arise in certain string-theoretic models as well as developing various teaching curricula with particular focus on involving undergraduate students in theoretical physics research. His future writing plans involve a technical text on certain aspects of the string theory landscape as well as an undergraduate textbook on spacetime physics in both special and general relativity.

- "*String Theory in the Classroom. A Case Study*". D. Cho is currently assistant professor of physics at Kenyon College. His area of specialty is gravitational physics. He has worked on a range of problems from gravitational radiation reaction to quantum gravity in higher dimensions. His recent research is focused on gravitational radiation reaction problems associated with rapidly rotating neutron stars, and quantum field theory in the early universe.

- "*Science Fiction of Everything*". Florian Conrady is currently a postdoctoral researcher at the Perimeter Institute for Theoretical Physics in Waterloo, Canada. He previously held a postdoctoral position at Penn State University and he received his doctorate from Humboldt University in Berlin. The main topics of his research are quantum theories of gravity and topology; in particular loop quantum gravity and spin foam models.

- "*A Vision of Quantum Gravity*". Tim Koslowski is currently a postdoctoral researcher with the quantum gravity group at the Perimeter Institute. He is working on various approaches to quantum gravity, in particular loop quantum gravity and the asymptotic safety scenario. He studied physics at The University of Texas at Austin and the University of Würzburg.

- "*Does History Repeat Itself in a Non-Cyclic Universe?*" Sundance Bilson-Thompson obtained his PhD in lattice quantum chromodynamics (QCD) from the University of Adelaide, followed by a year of postdoctoral work in Seoul, South Korea. Dr. Bilson-Thompson proposed a model of the substructure of quarks and leptons which turned out to have possible relevance to quantum gravity. This earned him a three year postdoctoral research position at the Perimeter Institute. In late 2009 he was awarded a postdoctoral research position at the University of Adelaide.

- "*Shooting in the Dark*". Sabine Hossenfelder is assistant professor of physics at the Nordic Institute for Theoretical Physics (NORDITA) in Stockholm, Sweden. She focuses on research in physics beyond the standard model and the phenomenology of quantum gravity. Together with her husband, Dr. Hossenfelder maintains a physics blog called "Backreaction" at (*http://backreaction.blogspot.com/*).

- "*Deciphering Quantum Theory*". P. Goyal is a theoretical physicist specializing in information physics and the foundations of quantum theory. His recent research focus has been on creating an information-theoretic underpinning for quantum theory. Currently a postdoctoral fellow at the Perimeter Institute, he was previously at the Cavendish Laboratory at the University of Cambridge, where he did his doctoral research and was subsequently a departmental research fellow. In September 2010, he will start a new position as assistant professor in information physics at the State University of New York at Albany.

I would like to thank all of the authors for pouring their minds and souls into these articles; sharing with the reader their views on this most exciting of times. Particular thanks and gratitude are due to Professor John Donoghue for writing the foreword. Such a contribution from a scientist of his caliber is an honor and a privilege. Many thanks are due to Dr. Tasneem Husain for agreeing to act as editor to my own contribution and providing valuable feedback. I am also very indebted to my friend Dr. Mohamed Anber for going through the manuscript in detail and proposing essential corrections. Last, but by no means least, I would like to take the liberty in representing all authors in deep thanks to all of our unsung heroes, those who are always the backbone of any effort: family members, spouses, children, mentors, friends and all significant others.

Manal and the rest of the M5: this is for you ...

Moataz H. Emam
Cortland, NY

LIST OF CONTRIBUTORS IN ORDER OF APPEARANCE

"Are we there yet? The search for a Theory of Everything"

Editor: Moataz H. Emam

Foreword by: John F. Donoghue

1. **John F. Donoghue** (The Foreword); professor of theoretical physics at the University of Massachusetts at Amherst.
2. **Tasneem Zehra Husain** (The Fire in the Equations); assistant professor at the LUMS School of Science & Engineering, Sector U, D.H.A, Lahore, Pakistan.
3. **Peter Woit** (String Theory and the Failure of Unification); senior lecturer in the mathematics department at Columbia University.
4. **Moataz H. Emam** (Can String Theory Survive Complete Falsification?); assistant professor of physics at the State University of New York College at Cortland.
5. **D. Cho** (String Theory in the Classroom. A Case Study); assistant professor of physics at Kenyon College.
6. **Florian Conrady** (Science Fiction of Everything); postdoctoral researcher at the Perimeter Institute for Theoretical Physics in Waterloo, Canada.
7. **Tim Koslowski** (A Vision of Quantum Gravity); postdoctoral researcher at the Perimeter Institute for Theoretical Physics in Waterloo, Canada.
8. **Sundance Bilson-Thompson** (Does History Repeat Itself in a Non-Cyclic Universe?); postdoctoral researcher at School of Chemistry and Physics, University of Adelaide, SA 5005, Australia.
9. **Sabine Hossenfelder** (Shooting in the Dark); assistant professor at the Nordic Institute for Theoretical Physics (NORDITA) in Stockholm, Sweden.
10. **P. Goyal** (Deciphering Quantum Theory); postdoctoral researcher at the Perimeter Institute for Theoretical Physics in Waterloo, Canada.

CHAPTER 1

The Fire in the Equations

Tasneem Zehra Husain

LUMS School of Science & Engineering, Sector U, D.H.A, Lahore, Pakistan

Abstract: The five parts of this essay are held together by the common theme of unification. Each section focuses on a particular theoretical development and is written in the voice of a scientist living at the time, reflecting on the discoveries of his or her age, what they mean, how they demand a changed worldview, and what might potentially lie in the future. The first section is written as a letter from a young British student at Cambridge, to his sister. He describes to her the wonders of Maxwell 's equations and what they mean, but in terms that his sister, who is an aspiring painter, would be able to relate to. The second section, about relativity, is written by a layman. A New York school teacher who admires Einstein has gone to Battery Park to join the crowds welcoming the Nobel Laureate to America. Whilst waiting for the ship, he thinks about the theories that made this man an international celebrity. The third section is based on the reflections of a young European researcher in Copenhagen in the early 1930's. He sees the greats of quantum mechanics walk the corridors and learns from their conversations, whether directed at him, or overheard. The basic developments are mentioned, including Heisenberg 's uncertainty principle, Pauli 's exclusion principle, Schrödinger 's equation and wave-particle duality. The fourth section is set in the early 1980's. A third-world scientist, early in his career, is visiting the International Center for Theoretical Physics at Trieste, Italy. In a letter to his young children, he talks about particle physics, introduces the Standard Model, and ruminates on how essential collaboration and a conducive atmosphere are for research. The last scientist is a female graduate student at Harvard in the present day. She is beginning to study string theory and is fascinated by the promise of ultimate unification. After a brief review of string theory – and a mention of the LHC – she addresses the question of whether or not she thinks this might be physics' last word.

MAXWELL: ELECTRICITY AND MAGNETISM ARE UNIFIED

Cambridge

Michaelmas 1899

My Dear Sister,

Before I begin I must beg your indulgence, for this letter will not be nearly as long as you might have wished. I have been meaning to write to you all week but by the time I come back to my rooms each night, I am so exhausted that the very thought of putting pen to paper is a fatiguing one and I resort to having conversations with you in my head as I drift off to sleep. Tonight, however, I am resolved to write. Before I put my pen down, you will know why.

Remember how, when we were children out for a walk in the fields outside our house, every now and then one of us would spy a treasure; a speckled blue egg perhaps, that carried within itself the promise of a sparrow, or jewel toned leaves, in whose crisscrossed veins we would read maps to buried pirate's chests? Whoever first spied these riches would rush excitedly to the other, impatient to multiply the joy by sharing it. Some of my fondest memories are of you as a little girl running across the grass crying "Look, look!" Well, this letter is my version of that cry.

After we got over the first happy shock of discovery, I would turn instinctively to books or pepper Mother and Father with questions about these mysteries; I had a burning need to know why things were so, while you would keep the objects in your hands, turning them over silently, until they turned into your wonderful paintings.

I hope you are still doing that. This might seem like a strange remark, quite disconnected from anything, but it really is not. You have an eye for color and form and your talent should not be dismissed merely as a pleasant pass-time for a young girl. It is a true gift, and the more time I spend at this University, the more firmly I am convinced that it is vital for one to pursue one's passion. While my way is often hailed as being more scholarly and serious, there is no reason why the Universe should whisper more through Science than through Art. Regardless of the path, strange and wonderful secrets creep out of the shadows for those who walk with wonder.

In fact, now that I am learning about them a little more, I cannot but wonder what deep truths the colors know; given the worshipful reverence you approach them with, they might well be induced to share some of their ancient lore with you. You will surely be curious as to what instigated these somewhat uncharacteristic reflections. Well, the fact is that I recently found out something absolutely marvelous. I learnt that when the red of fire, the blue of water and the green of earth, meet in equal measures and whirl, like mystic dervishes in Eastern lands, they throw off their individualities in a wild frenzy and merge into each other to become pure brilliant white. The instant I heard this, I thought of you.

Surprisingly enough, it was not an artist but a scientist by the name of James Clerk Maxwell who first witnessed this miracle. For part of his experiments on light and color, Maxwell spun brightly painted wooden tops and watched as the distinctions between the sharply defined colors blurred and melted before his very eyes, the trinity of red, blue and green went into a hypnotic trance and emerged at the other end, as *white* light.

It may be that a veil was drawn from his eyes in those magical moments, enabling him to spy an underlying unity in apparently disparate objects, for a few years later, Maxwell was able to see a beautiful and exceedingly harmonious relationship that tied together phenomena which were obviously distinct.

But I am getting ahead of myself. Luckily, in the three or so decades that have passed between then and now, Maxwell 's work has passed through the sieve of many discussions; concepts now sifted, are sharper and clearer. The equations Maxwell wrote down have been re-derived and cast in many different ways until their message was absorbed and is accessible now even to those who, like myself, cannot without help see the secrets in the shadows. Let me attempt to point out those nebulous forms to you too.

I have pestered you endlessly about electricity, so I will restrain myself from doing that again now. You already know about there being two types of electric charges, called by convention *positive* and *negative*, that like charges repel each other whereas unlike charges attract; I will spare you – for the moment – further reflections on the subject, but I cannot promise how long this resolve will last.

Remember the fond hours we spent with father's compass, provoking the hapless needle to turn now here, now there in an endless circular chase? You, I recall, were particularly fascinated by the fact that an object placed outside the compass could exert a pull on the needle without ever touching it. The only ways we knew to move things were to physically tug away or push with all our might – this mysterious force seemed wasted in the aimless wanderings of a needle when it would have been so much better employed in moving chairs or houses or trees or people … our plans for magnets were really quite elaborate!

After all these years, I think I finally understand the predicament of the poor compass needle. It must forever take its marching orders from the *magnetic field.* And it is quite the same for electric charges too; they are doomed to follow the whims and dictates of the *electric field.* But what is a field, you ask? The idea is that each electric charge carries with it an invisible net that it throws out into space (an analogous case can be made for magnetism, so I am going to spare you the repetition and will just talk about one phenomenon); this net, which is thickest close to the charge and gets progressively thinner as it moves outward, is called the *field of the charge.*

Now, truth be told, Maxwell did not quite conjure the idea of a field out of the blue but rather built upon a concept inherited from another brilliant mind; Sir Michael Faraday. This is not quite to the point, but the history is a rather romantic one and in fact reads almost like one of those fairy tales you always had your nose in, and so I cannot resist telling it to you.

A hundred or so years ago, a son was born to the family of a poor blacksmith in London. As soon as he was old enough, Michael, as they named him, had to help bring in some money, and so at age 14 he became apprentice to a book-binder. This good man, recognizing the glint in Michael's eyes, encouraged him to read all the books he wanted in the afterhours. And so it was that our unlikely hero was possessed by books; by the light of day, he would tend to them, healing them as it were and in the evenings, the grateful books would return the favor, pouring all they held into his ever expanding mind. Almost seven years passed thus, until one day a customer at the store took young Michael along to hear a lecture by Sir Humphrey Davy at the Royal Institution. This was Faraday 's equivalent of the Ball in your Cinderella story, for it was here that things magically transformed. Not only was his own mind set on fire by what he heard at the lecture, but somehow the great Sir Humphrey too was so struck by the boy's enthusiasm that he took him on as an assistant. Even though Faraday had little formal training, his years with the books paid off; he had a feeling for his subject and he soon became a very successful experimentalist. So much like you used to say: "it all worked out in the end".

Back now to my main story. It was Faraday who postulated the existence of what he called *lines of force* to explain how two magnets separated by a distance, could exert a force on each other. These lines, so distinctly visualizable, existed in the seemingly empty space between the two objects and they were what "mediated" the force. Maxwell realized the intrinsic elegance of the construction and marveled at how Faradays's "lines of force can weave a web across the sky and lead the stars in their courses without any immediate connection with the object of their attraction". However, where Faraday was relatively unskilled in mathematics, Maxwell was a veritable prodigy. He took Faraday's intuitive idea, clothed it in the appropriate mathematics and honed it. In fact, as Faraday later wrote to Maxwell "I was at first almost frightened when I saw such mathematical force made to bear upon the subject, and then wondered to see that the subject stood it so well". But rephrasing of the concept, as you will soon see, was not merely an empty academic flourish but actually a crucial step that lead eventually to a brilliant insight.

And thus the concept of *field* was born. But I still have not quite told you what a field is, have I? Well, to begin with, it is not a tangible thing, but instead an intricate lace of messages written in a language known only to electric charges. The messages are clear instructions that tell other charges how to move – exactly how far and in precisely which direction; these directives exist at *every* point in space, whether or not a charge is present to implement them. Any charge, when placed in a field, will read the directions inscribed at the point where it finds itself, and must act accordingly; and hence one charge seems to "push" another, even though the other is so far away. By a previously agreed-upon convention, the messages are addressed to a single positive charge. My guess is that these are the simplest creatures, unable to decode anything but the most direct instruction; all other charges must do some basic arithmetic to find out what instruction is meant for them. Say an object carries three positive charges, it must implement the basic message thrice over – since a direction can't be tripled, it merely moves three times as far, in the prescribed direction. Negative charges, however, are contrary creatures; while an object carrying five negative charges will travel five times as far as the basic message says, it will do so in *exactly* the opposite direction! Quite entertaining, is it not?

There is of course a complication I have omitted in this discussion; the fact that every new charge we bring in casts its *own* net of instructions! This net moves outward from the charge in all directions, spreading through space as a rippling wave, its strands interweaving seamlessly with those of fields already present. An innocent wayfarer who now wanders, deliberately or unwittingly into this silken web, must obey the dictates of the new field which combines decrees mandated by each of the charges ; of course the wayfarer comes enshrouded in a net of its own, … and so the situation can get infinitely complicated, as is true of any party where *every* guest insists on talking! Sometimes, just for ease of conversation, one talks of "test" particles which are charges that can hear the ambient field without feeling compelled to add their own voice to the hum; but these are just mental tools we use to simplify the discussion – no such blissfully silent charge actually exists.

Even though I have only mentioned electric fields thus far, the selfsame discussion holds for magnetic fields. That, in fact, is where your compass needle comes in; it too is a magnet and as such both shouts out its own magnetic field and keeps its ears peeled for the messages of others. In the absence of any cross-talk to the contrary, the needle pays heed to the magnetic field of the earth and aligns itself accordingly. If, however, there is another magnet closer at hand, sending out louder instructions, the faint shadowy calls of the True North are drowned out and the compass needle takes its marching orders from the this new Drill Sergeant.

This has been a long digression, I know, but as you will see, it is important that you understand what a field is. As a mathematical tool, the concept is an old one; part of Maxwell 's genius was that he applied the idea to understand both electricity and magnetism and combine them into what is now called *electromagnetism,* also known as *electrodynamics.* This truly was a brilliant stroke, because other than the fact that each allows two types of charges [1], and that these charges either attract or repel each other, according to the adage "opposites attract", there really is no apparent similarity between the two phenomena.

In fact, down the ages, electricity and magnetism were thought of as intrinsically different qualities, each as distinct from the other as red is from blue. I will not go into the laborious details as to how it happened, or this letter will become a tome, but gradually over this past century, a realization dawned that the distinction between electricity and magnetism is perhaps not quite as fundamental as we had thought. At the very least, it was clear that the one was not completely indifferent to the other, but in fact there was more; one even saw slowly mounting evidence that some trade did occur across the borders of these two countries ….

Compass needles, which should have been subject to the dictates of magnetic fields alone, were seen to deflect in the presence of wires through which electric currents flow. Upon further investigation, it was observed that, depending on the direction of the current flow, two such wires attracted or repelled each other just like magnets would! A decade or so of furious experimentation lead to the conclusion that this phenomenon could also be *reversed*; the honor of producing electricity *from* magnetism fell to our friend Faraday who demonstrated that a changing magnetic field induced a current to flow in a nearby circuit. The results had been tallied. It was obvious that we had stumbled upon some fundamental connection between electricity and magnetism. The precise interpretation however, was not quite so clear.

This was the scene when Maxwell stepped onto the stage. Invoking the field, much like magicians of old waved their magic wands; he suddenly made sense of everything. Faraday 's observations could be condensed into a neat little set of equations which laid out how an electric field can result from changes in a magnetic field. In effect, these equations sketch, in their own mathematical shorthand, the following picture. Consider for the moment a field that consists purely of messages addressed to magnets. If the message at every point is fixed, an electric charge that strays into the fold will be blissfully oblivious to all the instructions floating in the air around it. If, however, these messages change with time, the electric charge too, is affected. Rather a surprising discovery, this. If you are impervious to certain aspects of your surroundings, it would be quite logical to assume that you would be indifferent to how they change; not so electric charges. They completely ignore the hum of the constant *magnetic* messages in the background, regardless of how loud and strong these might be – but the instant there is a change, the electric charges prick their ears and take notice. As it turns out, there is no room for eavesdroppers in nature, and what one hears, one must obey; hence electric charges, who completely brush off the whispers of a constant magnetic field, find themselves constrained to obey certain dictates in a changing magnetic field. In fact, it is almost *as if* they are in the presence of an electric field.

What if, Maxwell said, the situation is more symmetric than we had realized? Perhaps a magnetic field can be brought into being not just by a current but by any changes in an electric field ! And so, he wrote down another little set of equations describing this process; the two sets of equations were almost mirror images of each other and when put together, they turned out to be far more potent than anyone, even Maxwell, had yet suspected!

With his equations captured on paper, and all this going on in the background of his mind, Maxwell then began to ponder what happens when the field's "instructions" encoded into a point are not static but in fact swing back and forth like a pendulum bob in a grandfather clock. In such a case, even standing still, one would hear constantly changing instructions; the field starts at a particular value and gets progressively stronger, demanding an increasingly dramatic response, until it finally builds up to a *crescendo* and then, exhausted, gets weaker and weaker until it returns to its initial value, only to start the journey again. Maxwell, studying such oscillating fields, came to a startling conclusion. Suppose you start with a magnetic field whose changes are no longer arbitrary or unbounded but in fact occur in the tightly controlled cycles I have just described. The sympathetic vibrations of an electric field are set up in response to the ebb and flow of a magnetic field, and so the circle keeps turning; oscillating electric fields come in from the void to answer the call of oscillating magnetic fields and like the unending reflections in a hall of mirrors, they bounce off each other, *ad infinitum.*

---

[1] In the case of magnetism, it is not true charges that are allowed, but rather the two poles of the magnet.

Etched into his equations, Maxwell found, were the ghosts of these untiring oscillations closing in perfectly time and again, only to move apart once more. The whole exercise seemed quite pointless … until suddenly, in a moment of blinding illumination, the subtlety of this dance dawned on him. In an unparalleled flash of insight, Maxwell realized that all he had seen so far were the seamless movements of the *corps de ballet* weaving back and forth on stage. As his inspired gaze now pierced the swirl, Maxwell spied the *prima donna* – hidden in the back of the equations, visible only after some mathematical manipulations, was a graceful tide of light blissfully engaged in invisible pirouettes.

In essence, what Maxwell realized was that just like an electric field would appear to answer the calls of a changing magnetic field, a magnetic field could also be birthed by a changing electric field; in fact, an oscillating electric field causes an oscillating magnetic field and these two cannot exist together without giving rise to an invisible third. There is an inevitability in how these events unfold. Embedded into Maxwell's theory of electromagnetism was a *wave*, a self-sustaining undulating disturbance in the electromagnetic field; this whisper not only spread through the field at the speed of light, Maxwell declared, it *was* light.

The electric and magnetic fields bobbing smoothly back and forth sketch out a square to call home; adopting complementary polarized roles, one of the fields moves relentlessly up and down to span the walls while the other moves diligently at right angles, left and right, to put down the roof and floor. The child of this gentle motion is also a wave, but a particularly restless one. At the very instant of birth, it breaks headfirst out of its planar home and races perpendicularly outward at the speed of light, into a direction uncharted by either of its parents. This wave, said Maxwell, is not merely a rebel imposter but in fact a trailblazing cavalier – it is light itself; and all that needs to be done to give it birth is to set into motion a system of oscillating electric fields. These will ward off their loneliness by calling into being oscillating magnetic fields who then return the favor by summoning to the fore oscillating electric fields – and in this eternal rocking back and forth is the cradle for the electromagnetic radiation known as *light.*

Such is the story behind the dim wavering light of the flame under which I write this letter, and the light – perhaps the bright straight beams of the sun – under which you will read it; wondrous, is it not? But even more wondrous, at least to me, is the fact that there is such a deep common thread running through such apparently diverse phenomena. Maxwell has, probably for the first time, come up with a scheme that subsumes two previously separate entities, uniting them into one coherent, consistent whole.

Now you see why I had to write to you, even in this state of near slumber? I might not be fully coherent, but just could not wait another few weeks to share these jewels. I knew that you, who are constantly looking for the common ground between personalities that are apparently incompatible, would find this idea oddly comforting. I have admired, always, your belief that if people could only realize all the things they had in common with each other, they would connect in a completely different, more intimate way – and that then there would be peace on earth and unity between Men. For my part, I hope your faith is justified, but personally speaking I am extremely intellectually intrigued by this concept and am feverishly eager to know if the scheme can be extended to include other natural phenomena as well.

I cannot wait to be back home for Christmas break. I am longing to be surrounded by family and to take long rambling walks with you. Perhaps together, we too, can pluck some invisible ideas from the "fields" outside our house. Who knows?

With Great Affection,

Your brother James.

In 1879, J.C. Maxwell died, passing the baton on to Albert Einstein, who was born the same year. Einstein, whose thoughts went into free fall and rode light. Einstein, who got rid of the omnipresent ether, enforced a cosmic speed limit, allowed particles to appear out of nowhere and solemnized the vows between space and time. Einstein, who took solid ground out from under our feet and placed us on a rubber membrane that shudders at our every move; as these ripples tremble through space, he said, they will share stories about their origin with all they encounter.

EINSTEIN: SPACE AND TIME ARE WEDDED. MATTER AND ENERGY DECLARED INTERCHANGEABLE. MATTER CURVES SPACETIME, AFFECTING ALL OBJECTS MOVING THEREIN

April 3rd, 1921

New York

It finally happened. He arrived yesterday. The Times screams it out in its headline this morning: "Prof Einstein Here. Explains Relativity. 'Poet in Science' Says It is a Theory of Space and Time. But It Baffles Reporters"

I hardly slept last night, so anxious was I for the morning paper; I wanted to read what people said about it, I wanted to see the photographs, I suppose what I really wanted was to relive the experience – for among those "thousands of spectators ... in Battery Park when the mayor and other dignitaries brought Einstein ashore on a police tugboat", I was also present.

I was surprised at how uncharacteristically assertive I was as I propelled my way to the front of the singing, flag-waving masses awaiting the arrival of the steamship Rotterdam. A few people glared at me as I pushed past them, but I was only dimly conscious of their disapproval. All I knew at that moment was that I had to see Einstein ; the man who has unknowingly been my invisible mentor this past year. How much he has come to mean to me!

I first heard of Einstein, together with the rest of the world, in November 1919 after the now-famous results of the Eddington expedition were announced in London. The President of the Royal Society declared Einstein's theory to be "one of the most momentous, if not *the* most momentous, pronouncements of human thought" and the Times of London reported the fact that there had been a "REVOLUTION IN SCIENCE. NEW THEORY OF THE UNIVERSE. NEWTONIAN IDEAS OVERTHROWN". But even after the bending of starlight had been witnessed, and Einstein's idea-world made manifest, the general feeling towards the theory still seemed to be one of confusion. The consensus was that no one had yet succeeded in stating it in clear language. The New York Times further strengthened this impression in its own breathless headline, which read: "LIGHTS ALL ASKEW IN THE HEAVENS. MEN OF SCIENCE MORE OR LESS AGOG OVER RESULTS OF ECLIPSE OBSERVATIONS. EINSTEIN THEORY TRIUMPS. STARS NOT WHERE THEY SEEMED OR WERE CALCULATED TO BE, BUT NOBODY NEED WORRY. A BOOK FOR 12 WISE MEN. NO MORE IN ALL THE WORLD COULD COMPREHEND IT, SAID EINSTEIN WHEN HIS DARING PUBLISHERS ACCEPTED IT."

My curiosity piqued, I decided to find out more. I discovered that this acclaim had not come overnight. In fact Einstein had already done his part of the work more than a decade ago. What we were celebrating was the fact that the rest of us, who do not see quite as he does, had finally witnessed the *proof* of his visions. One of the concrete predictions of *general relativity* was that light does not travel in perfectly straight lines where a gravitational field is present. The corresponding deflection of starlight close to the sun was measured during a solar eclipse in 1919, in precisely the degree predicted by Einstein.

It turned out that there was not just a general theory but also a special theory, which had come before it. And these theories, which question all our preconceptions and turn everyday logic on its head, were conjured up by the power of pure thought, by a clerk who worked by day in a Swiss patent office. Just the idea that such a thing is possible, was absolutely thrilling to me. Much as I love teaching my high-school students, it is not always satisfying. For years I had resented not having the money and the opportunity to go on to University, but after hearing about Einstein, I realized that my job does not have to define me.

Bernard Shaw said that where Napoleon and other great men made empires, Einstein had made a universe. I suppose that aspect of it is what really appealed to me; given my own meager station in life it is reassuring to think that brute power can subjugate only so much, whereas the dominion of the intellect, which can monopolize, is unbounded. The feeling was intensely liberating. I had long since yearned for the chance to flex my mental muscles with a real intellectual challenge and so I decided to try my hand at learning this theory. I looked around for a suitable text and found that a surprising amount has been written on the subject. However, realizing that I would in all probability not completely understand every detail, I decided to focus instead on learning how Einstein thinks. And so it was that I picked up his own book entitled simply, "Relativity".

As I stood at Battery Park, waiting for a glimpse of the man, I thought of how often I had heard his voice in my head. Reading his book had made Einstein come alive. The very first lines drew me in. "In your schooldays, most of you who read this book made acquaintance with the noble building of Euclid 's geometry, and you remember – perhaps with more respect than love – the magnificent structure, on the lofty staircase of which you were chased about for countless hours by conscientious teachers". He wrote so intimately, I immediately settled in to what felt like a conversation with a friend, not a lecture by a "scientific mind whose deductions have staggered the ablest intellects of Europe", as the newspapers proclaimed.

Fascinated, I tagged along as Einstein walked me through the concept of truth (as being something that follows logically and consistently from a set of axioms) and convinced me that to talk of bodies changing their position in space with time, without "serious reflection and detailed explanation" would be tantamount to "loading my conscience with grave sins", since "it is not clear what is to be understood by *position* and *space*". Thus gently, he nudged me to probe the calm waters of the seemingly simplest concepts and examine the many layers of unstated assumptions that lay bubbling underneath.

In the very early years of this century, when Einstein turned deep conundrums over in his head on his walk back home through the picturesque streets of Bern, light was very much on his mind. Light, in fact, had turned out to be quite a problem. Not content, apparently, with the love and attention Maxwell had showered on it, light began to upset the old guard; it came up against and challenged concepts as deeply held and cherished as the laws of Galileo and Newton.

Centuries ago, Newton laid down a law of physics, clearly and firmly. In a decree I pass down to my own students every year, he ordained that "Every body will continue in a state of rest, or of uniform motion in a straight line, unless compelled by an external force to do otherwise"; this, Newton's first law of three, is known as the *law of inertia.* Inertia is the tendency of a body to resist change, to continue in the future as it has in the past; once placed somewhere, it tends to stay put; if in motion, it tends to keep going, at the same pace, in the same direction – as long as it is left alone. Consequently, those of us who observe the world from platforms that are either uniformly moving, or still, are called *inertial observers.*

Newton argued that two inertial observers should perceive the laws of physics as being the same. In other words, nature should not appear to be fundamentally different to me, standing here, and you, standing five feet to my left. If both of us throw identical balls in the air, at exactly the same speeds, they will land identical distances away from us. Nor should we disagree if one of us hops on a railway carriage that glides smoothly along the tracks, never speeding up, or turning, or slowing down. Indeed, if the carriage had no windows to the outside, and one could not see the landscape whizzing by, one might never realize that the train was moving at all. So, if an inertial observer can never say for sure whether or not he or she is moving, clearly the basic laws of physics must remain unchanged under uniform constant motion. Thus Newton declared, and since this law was eminently reasonable, thus was the word passed down through the ages. This statement is called the principle of (Galilean) relativity and was held to be so obvious as to not require any justification. The arrangement worked well for centuries, for as long as one expected all natural phenomena to be explicable using classical mechanics, but when Maxwell 's equations came along and elucidated the nature of light, it became clear that light refused to comply with the age old principle.

What played out in Einstein 's head was nothing less than the epic confrontation between Maxwell 's new but spectacularly successful theory, and the long revered and obviously correct framework of classical mechanics. When the theories were pitted against each other, the battle was subtle, but fatal. As these two sets of equations met in head to head combat, many long held, deeply rooted beliefs went down in flames; out of these ashes, rose the phoenix of *special relativity.*

While I can not at all claim to understand the mathematical derivation of the theory, the intuitive idea does make sense when he reasons it out. I suppose it is accessible because Einstein thinks in pictures, not equations. Those last he saves for later, to work out exact relationships and frame his ideas more precisely, but the flash of genius is in the mental image, not the diligence of his mathematical manipulations. His ideas, though revolutionary, can thus be visualized; one can conceive of the problem and marvel at its extremely unexpected resolution.

Einstein dealt with the abstract conceptual clash between electrodynamics and Galilean relativity by first of all making it concrete. He did this by the prosaic instrument of a railway carriage which he imagined to be traveling along the tracks at a constant speed and direction, changing its position with respect to the stationary platform, yet doing so without rotating. I have become rather attached to this hypothetical carriage, so often does it appear throughout his book, so often have I rode it in my mind, chasing after the arguments Einstein makes, attempting to catch up … I feel quite a sense of ownership now with what Einstein calls "our old friend the railway carriage". To complete the setup of our mental experiment, we place a man on this carriage (call him Albert) while his brother (whom we name Marcel), stands on the platform; being inertial observers, both see natural phenomena run their course according to the same general laws. While the actual values they assign to quantities such as velocity and distance might differ, the brothers agree on the laws of physics.

Now, just to get our feet wet and to acquaint us with the way things work, Einstein takes a tiny familiar step. Imagine a raven, he says. I found this quite reassuring; there is nothing remotely scary or mathematical about a raven. This particular bird we set flying through the air in such a manner that its motion is uniform and perfectly parallel to the railway carriage. As is expected, Marcel will see the raven as flying faster than does Albert; this must be the case, since the latter is himself moving in the same direction as the raven, thus decreasing the distance between them. Moreover, if Albert sits perfectly still, he is not covering any distance in his own *frame of reference,* which is just a mathematical way of saying "point of view"; as he sees the world, he is not moving. According to Marcel, however, he is moving further and further away, covering (the same) definite distance with each passing moment, simply by virtue of the fact that the railway carriage is moving.

We would expect there to be a dictionary that allows us to take the measurements made by one brother and convert them into those of the other, and indeed there is - simple addition. It clarified things for me to assign values to the speeds, so I sped my mental train along the tracks at 50 km/s and let the raven fly so that its speed, as measured by Marcel, is 70 km/s. It is trivial to see that since in any one second, the raven has covered a distance of 70 km and the train has moved ahead by only 50 km, the distance between them is 20 km. Thus Albert measures the speed of the raven to be 20 km/s *in his frame of reference.*

So far there were no surprises and I felt myself settling comfortably into the book. Perhaps a little too early. So much for ravens, Einstein says, what now of light ? Suppose that Marcel shoots a ray of light in the direction of the train[2]. Clearly, he himself will see light travel at the speed c, while Albert, using our reasoning above, should see light travel at the speed 50 km/s less than c. But that can simply not be true. Light *always* travels at the speed c (roughly 300,000 km/s in vacuum); this is a general law of nature, derivable from Maxwell 's equations, and like every other general law, must hold in every inertial frame of reference.

Suddenly, we find ourselves up against a wall. Our familiar logic does not apply to light. Something is clearly wrong; either Maxwell 's equations are lacking, or else there is something amiss with the very sensible laws of Newtonian mechanics. Faced with this choice, the traditional approach might have been to reject the new and hold on to laws which had been borne out by years of everyday experience. Einstein, however, bet on Maxwell. This was not a rebellious contrarian choice, but one motivated by the depth and beauty of these breathtakingly symmetric equations which explained not only how things happen, but also why. Maxwell's theory was aesthetically and logically far more complete than Newton 's laws of classical mechanics, which, though monumentally successful were in effect just a working prescription.

Since he was questioning age old beliefs, Einstein had to build a new vision of the world from the ground up, based purely on logic. But even so, he needed a place to start. Some axioms, some *initial conditions* were needed. Einstein's brilliance lay in the fact that he took the two main elements of these seemingly incompatible theories and pieced them together into a new whole. As his postulates, he chose firstly that the laws of physics should appear the same to all inertial observers, and secondly, the fact (confirmed by the recent experiment of Michelson and Morley) that the speed of light is always constant appears to be invariant, regardless of the motion of the observer. These were the only concrete footholds Einstein had. Making no additional assumptions, he set about examining what had

---

[2] From now on, we assume for simplicity that the entire arrangement takes place within a vacuum where the speed of light is at its maximum. In other media, though still fixed and calculable from Maxwell 's equations, the speed is slightly slower.

to be true for these two statements to be consistent. Armed only with pure logic and deduction, setting all preconceptions aside, Einstein forged bravely ahead. Velocity was the key; velocity, after all, is what distinguishes inertial frames and it is through velocity that we link one such frame to another. In a brilliantly inspired move, Einstein started investigating the seemingly obvious, namely the way in which velocity is measured. Velocity is a composite concept, it is a ratio between distance and time; truth be told, these are the quantities we actually measure. In order to calculate the velocity of an object, we simply measure the distance it travels and divide this by the time taken for the motion. So, to reconcile the problems with the speed of light, Einstein returned to first principles and thought the scenario through, not skipping even the obvious steps, as I keep reminding my students not to do when they learn concepts for the first time; careless mistakes are made that way, I keep telling them. Just how consequential these mistakes can be, I myself had no idea until I let Einstein teach me how to measure velocities.

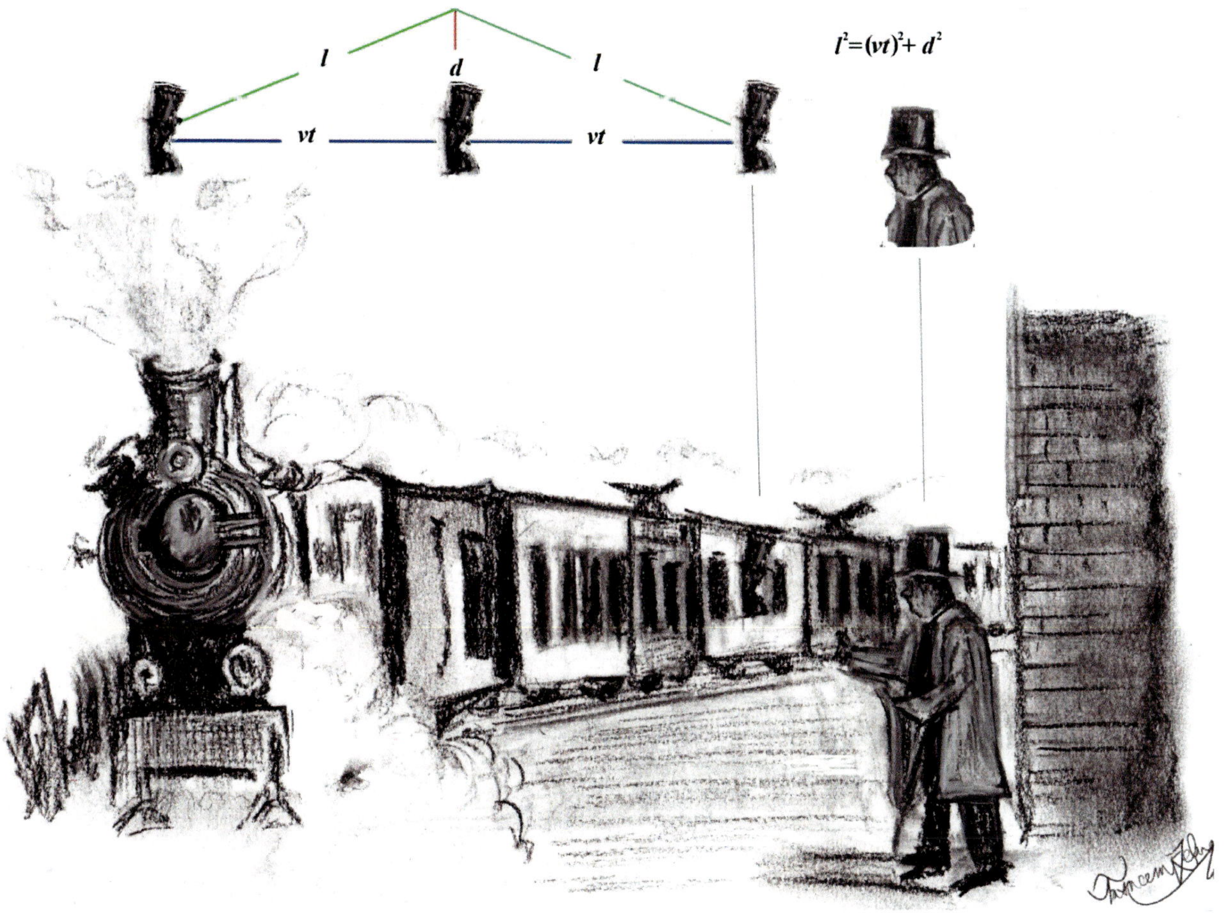

Figure 1: Albert, Marcel and the relativistic train. The formula shows that the distance that light has traveled from Albert's perspective (*d* – the red line) is shorter than that seen by Marcel (*l* – the green line). The distance the train traveled is *vt*; the speed of the train multiplied by the time it takes the light beam to cover the distance.

He takes us back to our familiar set-up of the railway carriage and breaks down exactly how each brother perceives the motion of the other. Suppose Albert aims a ray of light at the window, at right angles to the motion of the train – see (Fig. **1**). He can measure the distance between himself and the window and equally, he can measure the time the light takes to hit the window and be reflected back to him. Dividing the one by the other should give him the speed of light, the absolute value of which is fixed and known. Now shift perspective and step into Marcel's shoes. He too, sees a ray of light shot at the window but he also sees an additional effect; he witnesses the motion of the train, and as a result, the change in position of his brother. He knows that if the ray of light is to make its way back to Albert, it cannot simply head back in a straight line to where it came from, it must instead travel at an angle, aiming ahead, to get to where he now sits, waiting for the ray to return. In doing so, light effectively travels a greater distance than it appears to the unsuspecting Albert. Marcel is quite clear about this. He measures this distance, which turns out to be nothing more complicated than the hypotenuse of a right angled triangle (the sides of which are given by the distance between Albert and the window, and the distance covered by the train in the time taken for the ray of light to return. Now, since he is an inertial observer, who can claim full validity for his point of view, it only makes sense that he too, calculates the correct value for the speed of light.

How is this possible?

As is familiar to school children everywhere, it is not difficult to manipulate your way to the answer when you already know what it is! This is almost exactly what Einstein did. In a move of breathtaking audacity and simplicity, he said that the only way for both brothers to arrive at a consensus regarding the speed of light, is if Marcel – who sees the ray of light as having traveled a greater distance – sees this process take place over a *longer* time interval. In other words, Marcel will see clocks on the train as *running in slow motion*! This, somewhat unexpected resolution of the conundrum, turns out to be just what the doctor ordered.

Thus, said Einstein, time *appears* to be dilated, when an object is in motion. Since no one is ever in motion with respect to themselves, this only holds for other people's perceptions of my motion, or my perception of theirs. In my own frame of reference, things go along just as they always have. But, if I look at someone or something that is in motion with respect to where I stand, I will see their time as passing slower than my own. The paradox, of course, is that since they would be equally justified in claiming themselves to be still while I am in motion, they will see their own clocks as ticking away as usual, whereas *mine* is the one that has slowed down! There seems to be no way of reconciling this difference, no objective way of deciding who is right and who is wrong; both are equally justified in their claims. This conclusion is completely mind boggling, even to those of us who think they have followed the logical steps leading to it, but it is inescapable.

Through a series of elegant thought experiments, Einstein showed that events considered simultaneous by one observer will occur at different times as far as another observer is concerned. There is no notion of universal simultaneity, my past might be your future and vice versa. The arrow of time, which had marched on at the same pace eternally, now was free to speed up, slow down, and wind around. Apparently, there is no absolute concept of time; there is no universal clock out there in the sky, to which we all must tune our watches. We keep our own times, march to our own rhythms. It is a wonder to me why this liberation, this freedom to choose our own point of view and claim it as valid, should be so profoundly disturbing to Man.

Moving on from here, it was more or less clear that if motion made our perception of time blurry, it should do the same for our perception of distance; after all, the sacred ratio of the speed of light must be maintained. It turns out that following the same logic, one can – or at least Einstein can – argue that we perceive distances as shorter when they are in motion with respect to our view of the world, as opposed to when they are still. So, as if the mutiny against time was not enough, Einstein went on to declare that the notion of spatial distance was not sacred either. In fact, lengths seem to contract, when objects are in motion.

Einstein was able to show that rods moving relative to us would appear contracted in the direction of the motion and clocks moving relative to us would appear to run slow. In both cases, the effects would be almost negligible at the velocities we consider on an everyday basis, but would become increasingly significant as we moved faster, approaching the speed of light. In fact, at the speed of light itself, time would stop altogether and distances would contract to nothing, indicating that the speed of light had to define the speed limit for all matter in the universe. The speed of light is like the pot of gold at the end of the rainbow; no matter how close you come to it, the "end of the rainbow" moves further and the mysterious crock of Elven gold remains ever elusive. Matter, solid stodgy matter, turns out to be far more steeped in the Romantic tradition than one might expect, for it chases this unattainable goal in a quest that is doomed to fail.

This chase is not without peril either. It turns out that light is jealous of its throne and guards it perfectly; the speed it chooses to move at is exclusive, an option for it alone; all material objects must confine themselves to moving slower. The mechanism for enforcing this decree is rather clever and quite simple: the closer a massive object attempts to get to the speed of light, the heavier it becomes. All the energy channeled into making it moving faster goes mostly into making it heavier and heavier. Using comparatively simple mathematics, Einstein was able to work out that no material object can approach the speed of light, unless it has access to infinite energy; in other words, the feat is simply impossible.

Previously, the distinction in classical mechanics had been rather clear. There was Matter and there was Energy, both of which could exist in several forms (solid, liquid, gas; kinetic energy, sound energy, heat energy etc) and both of which were subject to *conservation laws*. The total amounts of matter and energy in the universe were thought to be fixed and all interactions had to be such that they honored this law. Einstein, however, showed that another kind of interaction was possible. Closing the centuries old chasm between matter and energy, he paved the way for matter to not only change form, but also to cross the bridge and change into energy, just as energy was now allowed to change into mass. Matter, he said, is just energy made manifest. It is in fact an extremely efficient storehouse, since unimaginably huge amounts of energy can be compressed into tiny pieces of matter[3].

In Battery Park yesterday, there was a sea of heads as far as the eye could see. Some were little copper heads gleaming in the sun, others had shiny black top hats on them, yet others had soft brown hair done up in curls; inside my own head of somewhat thinning black hair, I was turning over the pages of a beloved book when suddenly I heard an outcry. Ripples of excitement ran through the crowd, people started shouting, "He's here, He's here". Parents lifted children to their shoulders so that they could catch a glimpse of the great man. There was pushing and pulling all around me, but more stubbornly than I have ever done anything before, I dug my heels in and held my ground.

The hours of waiting had not sufficiently prepared me for the deep thrill I felt when I finally saw a man in a faded gray raincoat. His flopping black felt hat nearly concealing the gray hair that straggled over his ears, he stood timidly, facing a veritable army of cameramen. Through the shouts and shrieks of the crowd, the incessant questions of the reporters and the constant buzz of clicking cameras, he stood, clutching a shiny briar pipe in one hand and with the other, clinging to a violin. The woman next to me turned to her husband in amazement "He looks like an artist … or a musician" she said. I could not help the smile that floated to my lips. Little did she know, he was.

A chill went down my spine when I realized that Einstein was gazing in my direction. I felt my forehead wrinkle with intensity, as it always does when I concentrate. *Please see me*, *please see me*, I prayed. His eyes swept the crowd and just for an instant, locked with mine. He smiled and nodded oh so slightly. All around me, people were moving but I stayed frozen; perhaps in that long moment, I experienced time dilation, for I could have sworn I felt the march of time slow down.

By the time I came back to my surroundings and the movie reel of life began to move again at the brisk pace which felt familiar, cameras had stopped clicking. Einstein 's motorcade had apparently moved on and the crowd, having witnessed history, was in a rush to disperse. The people in my immediate vicinity were giving me wary sidelong glances, wondering if I would suddenly start pushing my way through again, as I had earlier in my haste to make it to the front. But I stood quietly, content to let the waves of the crowd break and wash all over me. I was in no hurry to get home, in no rush to break the spell of the moment.

And so I began winding my way home slowly, taking care to avoid the main streets on the Lower East Side. I knew, as did the rest of the city, that Einstein was headed for the Hotel Commodore and thousands of people lined the path, in a street party of sorts, waving to his motorcade. Even avoiding his route, I heard for a long time the honking of car horns; this modern form of fanfare being the tribute the streets of New York paid to the gentle genius.

How does he put up with this constant rush of adoration around him? I wondered. Perhaps these are the times he retreats to that secret chest in his mind; that room he imagined, so far out in dark empty space so as to be completely free of the gravitational pulls of any massive object; the room he invented when he came to grips with *general relativity.*

For, of course, despite the phenomenal success of the special theory of relativity, there was one obvious limitation. By construction, the theory dealt only with objects that felt no external force – in particular, they were sufficiently

[3] This is manifest in what is perhaps the most famous equation in the world: $E = mc^2$; which states that an amount of mass m is equivalent to an amount of energy E when multiplied by the speed of light squared.

far removed from any massive object so as to not experience any gravitational pull – and thus moved either uniformly in a straight line or remained at a state of rest. For Einstein, this limitation was a problem. "No person whose mode of thought is logical can rest satisfied with this condition of things" he wrote. In order to extend the validity of his theory, it was clear that he would need to incorporate gravity. But that involved first understanding what gravity is, so Einstein rounded up the facts as they existed at the time.

Ever since Newton got in the way of an apple's descent to the ground, it has been known that gravity is an attractive force that exists between any two massive objects. It is what keeps us anchored to the ground, just as it keeps the Earth pinned to its orbit around the sun. In fact, as Newton told the story, gravity is capable not only of acting over large distances but also of traversing these distances instantaneously. Moreover, another defining characteristic of gravity is that it causes all massive objects to accelerate indiscriminately, at the same rate. This last fact was allegedly demonstrated rather dramatically by Galileo who threw two stones of very different masses down from a great height at the leaning tower of Pisa, only to find that they reached the ground at precisely the same time. The only force acting on these balls, compelling them to fall, is their own weight[4], so if they reach the earth in a single thud, it must be that the acceleration due to their (different) weights is equal!

Having assembled all the facts, Einstein found that Newton 's tale was somewhat one-dimensional. The entire narrative read as a list of events, of causes and effects – if this happens, this shall follow. While the predictions tallied well with reality, there was something missing. One could read the entire story Newton penned, from beginning to end, and still not have a feeling for the character of gravity ; what is it really all about? What are its secret motivations for acting as it does? What are its demons – realms into which it does not venture? What unsuspected tricks can it perform? Einstein knew that before he attempted a retelling story, he had to ask some deep questions which would enable him to understand both the plot and the protagonist better.

He started by focusing on the gaping holes in Newton 's narrative. First, there was the part about gravity acting mysteriously over large distances. This *action at a distance* had already been dealt with beautifully by Faraday who invoked the concept of a field to explain a similar behavior in the electromagnetic force. Sensing both the beauty and the merit of this construction, and feeling no need to reinvent the wheel, Einstein extended the same mechanism to gravity. "Just as in electromagnetism, we think of a magnet attracting a piece of iron not directly but by calling into being something physically real in the space around it," Einstein wrote, we can think of two massive objects attracting each other across space through the medium of a *gravitational field.*

There was however, another problem with gravity. According to special relativity, there was no room in the universe for anything to break the definite speed limit imposed by light, not even gravity. Since instantaneous transfer of force was an intrinsic feature of Newton 's theory, it was clear that a new description had to be found.

Einstein thought and thought and thought. For years he grappled with this problem, and then one day, in 1907, Einstein had what he later called the happiest thought of his life. While sitting in a chair in the patent office in Bern, all of a sudden it occurred to him that if a person falls freely (under the sole influence of gravity and in the absence of any other force) he will not feel his own weight. "This simple thought made a deep impression on me", Einstein wrote, "It impelled me towards a theory of gravitation."

Admittedly, *this simple thought* did not do quite the same to me. Thus far, I was with Einstein, on the same page as it were – but in that sentence he lost me. I backed up and thought it through again, slowly. A distant memory flashed into view. When I was growing up, a group of the boys from my neighborhood would gather every evening at the empty lot in the corner and get up to all those sorts of antics which resulted in them being called *the hooligans*. They didn't do anything really wrong, but they didn't seem to do anything very right either. I could never have been a part of them, my mother would have had a stroke and besides, I wasn't exactly a daredevil and would not have fit in. But when I passed by the lot, I always slowed down a little and gazed out of the corner of my eye at this reckless world full of laughter.

Once, however, for a fleeting moment, I saw it from the inside. I was walking back home after piano lessons (my parents were convinced that someone who had as undoubtedly bright a future as myself, should be cultured in every

[4] For the purposes of this discussion, the effects of air resistance are comparatively small and can hence be safely neglected.

way). Much as I love music, my fingers refused to play the melody in my head and the lessons were extremely frustrating. My head hung down, I had almost passed the lot when a voice pierced my thoughts. "Jacob. Hey Jacob! Come here." I turned around, stunned, to find Isaac calling out to me. Isaac was a class-mate but not a friend and I still do not know what possessed him to call out to me then. It turned out that on that particular evening, the hooligans were attempting to fly. This involved scaling a broken wall, about eight or nine feet high, and jumping down. Isaac said it was quite the thrill and asked if I wanted to try. I was scared out of my wits, but also extremely excited and somewhat flattered at having been included. Yes, I said, of course.

Somewhat in a blur, I climbed the wall with an army of boys clapping and hooting and cheering me on. Close your eyes on the way down, they shouted. Just close your eyes and jump. So I did. It lasted less than half a second (I worked it out from the relevant equations when I got home) but in that brief instant, I really did feel free ... until of course my feet hit the ground. Thinking back on that now, I realize that the first time I felt a force was the impact of my landing. The danger of the jump was not in the process but in its ending.

A feeling of comprehension slowly creeps over me and I think quietly so as to not scare it away. While one is falling, one does not experience any force. Also, we know that any two falling objects experience the same acceleration; what does this mean? Well, say there are two boys who jump down from the wall at the same time. Since they start from rest and experience the same acceleration, they will, at each instant have the same velocity. This velocity will of course keep increasing the longer they fall, but it will at all times be the same for them both and so, my mind now races to the thrilling conclusion: they will appear to each other to be at rest! From here, it was quite simple to convince myself that any two objects in free fall will appear *to each other* to either be at rest, or undergoing *uniform motion in a straight line* (the latter being the case when, for example, one object has been in free fall longer than the other and thus has a *head start* on the velocity).

Although a little breathless, I have finally caught up to the page where Einstein sits patiently, waiting for me. He continues: It is settled that when you fall in a gravitational field, you feel no force and that an object falling with you will appear to be at rest. So then gravity is an agent that acts on every massive body just as hard as it needs to in order to make it accelerate at the predetermined rate; while the actual gravitational force changes for each object, the acceleration is a sacred constant. Hidden here somewhere, there seemed to be a deep connection between the nature of gravity and acceleration. Einstein was determined to hunt it down.

This is when he invented that chest in space, the one which was large enough to be a room and yet had no peepholes to the outside. As long as the chest is far away from any gravitational field, Albert, who lives inside it, is (with respect to Marcel who stands and watches from a distance) an inertial observer who goes about his business unfettered, as inertial observers do. Since he is not subject to any forces, Albert is free to float in mid air and stay perfectly still. If, however, he exerts a force, say by pushing against a wall, the reaction of the wall will impel him into motion; he will drift around in that room, colliding, perhaps with other objects similarly floating, any such impact resulting in each of them modifying their motion just as if they were so many billiard balls on a table.

This scenario seems to make sense, so we take it one step further. Suppose now that the entire chest is accelerated upwards at a constant rate. If Albert is standing on the base of the chest, he too is accelerated as a result of this contact; in reaction, his feet push down on the floor and he experiences the familiar sensation of weight. But what of an object that is not in contact with the chest and thus, does not feel the acceleration? What if Albert, used to living in the absence of forces, places an apple in the air, expecting it to stay perfectly still until he reaches out for it again? Now that the chest is accelerating, this is not quite what happens. From Marcel's point of view, the apple does stay suspended in space, but the chest, on its accelerated path upwards, "rises" to meet the apple, taking Albert along for the ride. As Albert sees it, however, the apple falls to the floor, accelerating at a fixed rate the entire way; this, of course can be very satisfactorily explained by assuming that he is at rest in a constant gravitational field.

This being the case, Einstein began to wonder if perhaps the effects of (uniform) acceleration are truly indistinguishable from those of gravity. Working his way through a few more thought-experiments, he eventually convinced himself that all accelerating observers can claim to be at rest, if they pin the blame of their observations on a suitable gravitational field. After all, if one cannot say for sure whether one is accelerating or experiencing a gravitational field, then the two situations, being interchangeable and indistinguishable, must be equivalent. But, said Einstein, if gravity can always be *explained away*, perhaps it is not really a force at all.

By the time I reached this point, I had quite the headache. In fact it was almost bordering on vertigo. These ideas were dizzying in their intensity and I could feel the pace building up; the crescendo was close now. I had to put the book down for a while, breathed deeply and let these ideas soak in before I loaded my brain with others.

Part of the problem with Einstein is that he does not ask questions one can walk away from. He will not say, as scientists are supposed to, "What happens if I change this equation around and add a term here?" but instead shows you an irresistibly vivid image. "What would I see if I rode light ?" he asks, or "What would I feel if I fell off a roof?" His way of phrasing problems, indeed his way of thinking, pulls you in and before you know it, you are standing at the mouth of a labyrinth.

Through his inspired logic, he leads you patiently through obstacles you really could not have navigated alone and then suddenly, after some unexpected turns and near misses, you find that you are again out in the sun, but in a world quite different from the one you left behind. It is hard to find fault with the steps once one has gone through them, but the overwhelming feeling is that one could not have done it alone. Yes there is logic in the process, but there is also a sense that the process was not discovered by pure logic alone; Einstein probably did have what has been called "a speculative imagination so vast that it senses great natural laws long before the reasoning faculty grasps and defines them".

Yesterday, when the distinguished guests arrived at Battery Park, reporters were, of course, asking questions of everyone in the party. To Weizmann (Einstein 's friend and traveling companion) they addressed the inevitable good-humored question: Did he understand his friend's theory? Weizmann replied "Einstein explained the theory to me every day on the way over and by the time we arrived I was fully convinced *he* really understands it". The crowd broke into laughter, but I kept chuckling even after the others had stopped. It was reassuring to hear that even the privileged few who were granted such intimate access to Einstein did not find the theory simple to understand. Suddenly my sleepless nights and throbbing head were vindicated.

But back now, to gravity. If it is not a force like other forces, Einstein said, and objects in the presence of a gravitational field can be thought of as being accelerated, we need to examine more closely how acceleration affects objects and their motion.

And so with this aim in mind, we return to that windowless room out in space, roll up our sleeves and get to work. Suppose a ray of light is shot from one end of the room to the other. Meanwhile, in the time it takes for light to cover this distance, the entire room is moving upwards, constantly gaining speed. When the beam actually arrives at the other end, the floor has "risen" from where it was. The point at which light hits this wall is considerably closer to the floor than the corresponding point on the opposite wall where the light started out. Moreover, as the astonished Albert realizes, this path taken by light is not just tilted downwards, it is *curved.* This result was exceedingly strange. Light travels in straight lines. Somehow acceleration was playing havoc with this long revered fact.

Not that the parabolic motion in itself was strange. Had it been an apple thrown outward instead of light, the outcome would not at all have been surprising. Falling objects are well known to trace curved parabolic paths in their descent to the ground. Galileo, in fact, had already worked out the equations for this projectile motion, which was used to great effect in cannons and other instruments of warfare. High school students today routinely work out problems showing how objects that are imparted only a horizontal velocity (balls thrown straight outward, for example) nevertheless develop a downward velocity due to gravity ; for objects in free fall, the horizontal velocity stays constant but the vertical velocity keeps increasing under gravity's accelerating influence and thus the paths traced out by them are not straight lines but curves.

This argument however, was supposed to apply strictly to material objects, to masses that could justifiably be said to succumb to gravity 's attraction. Light, being massless, was thought to be exempt from feeling, or participating in, this interaction. It was clear to Einstein that light would appear to be bent in an accelerating frame; the problem was that if gravity and acceleration truly were equivalent, he would have to find a way to explain this bending of light in a gravitational context also. In order to understand just how the trouble crept in, Einstein devised yet another mental apparatus to put the idea to the test. This time, it was a spinning disc.

People have remarked that Einstein 's uncanny ability to visualize these scenarios so vividly might, in part, be due to his thorough examination of patent applications. Einstein's phenomenal ability to conjure up imagined machinery must at least have been strengthened by the hours he spent pouring over a vast range of technical designs, checking to see whether or not they would work! But in any case, regardless of how it was thought of in the first place, there was now a disc in my mind. A perfect circle, with a radius 1 meter long, it rotates at a constant speed. I place two men on this disc; Marcel at the center, and the ever adventurous Albert out on the edge.

Sitting at the center, Marcel is still even while the entire disc spins. Albert however is being accelerated. Despite the fact that his speed is constant, his direction constantly changes and so, as far as Marcel is concerned, Albert no longer falls under the category of inertial observers who as we know must be in "uniform motion in a fixed direction". As far as Albert himself is concerned, he cannot *see* this acceleration but he can definitely feel a force, emanating as if from the center of the disc, pushing him out. This *fictitious force*, as it is known, has no external reality – it is merely an artifact of motion which disappears just as soon as the disc stops spinning. Be that as it may, while the disc is rotating, Albert experiences a force that appears to act so that all material objects are accelerated equally; he is thus perfectly within his rights to claim that what he feels is gravity.

Given these split yet equally valid perspectives, anything Albert sees is doubly important to us. We need to account for all his observations twice; once, as the experiences of an accelerating observer – as Marcel insists he is - and then again as the experiences of an observer at rest in a gravitational field, as Albert appears to himself to be. In order to have a measurement to interpret and compare, we ask both men to calculate the circumference of this spinning disc on which they sit. Albert immediately takes out a set of rulers and walks along the disc placing them end to end until he returns to his starting point. When he tallies the count, he finds that the circumference comes in as being exactly π. Albert knows that his is clearly the correct answer. From Euclid onwards, this has been one of the cornerstone results of geometry. A circle whose radius is one unit must have a circumference that measures π units around; this, in fact, could even serve as the definition of a circle, so fundamental a result is it. While Albert sits smugly, Marcel is rather worried. Try as he might, he cannot get the answer to work out as it should. Sitting at the center, he measures the circumference by counting the number of rulers needed to line the edge of the disc. But, since these rulers are placed at tangents to the outer rim, they are affected by the motion; to Marcel, each such ruler appears shortened and thus more of them are needed to go all the way round. Marcel's measurement of the circumference thus turns out to be *more* than π!

The only way the familiar laws of geometry could be recovered is if the two men measure different values for the radius of the disc as well; but this turns out not to be the case. The radius is perpendicular to the direction of motion and thus remains unaffected by the spinning of the disc. Both men concur – the radius of the disc measures exactly 1 meter. The inescapable conclusion stares us in the face. Ridiculous as it may seem, something is happening to geometry itself! Acceleration appears to *modify geometry* somehow, and if acceleration has this effect well then it stands to reason that gravitation must also.

From here onwards, mathematics plays an increasingly important role in the formulation of the theory. Moreover, the relevant mathematics is so complicated that even Einstein needed mathematician friends to help him out. I cannot claim to follow the details anymore, but the general idea makes sense.

As his friends explained, Einstein needed to study the geometry of curved surfaces. Euclid 's axioms, those tenets we use as pillars in most of our construction, apply only to flat surfaces – like a book, or a table; they do not hold for surfaces that are rounded or undulating. The rules of geometry are different for sheets and spheres. If circles with equal radii are drawn on the inside of a large round bowl, a flat sheet of paper, and the surface of a ball, the circumference of the first will be larger than that of the second, and that of the third will be smaller still. Other familiar rules change as well. The sum of the interior angles of a triangle is not always 180 degrees, contrary to what we have been taught all these years; that particular truth holds only with the caveat that the triangle is drawn on a flat surface. If it is drawn on a ball, the sum of its angles is greater than 180. And if all this wasn't enough, it turns out that a straight line is no longer necessarily the shortest distance between two points!

This last statement is perhaps the easiest to visualize. Think, for instance, of a circle, and mark off any two points on it – call them *A* and *B*. Starting from *A*, there are now two distinct ways of getting to *B*. We can travel clockwise, or anti-clockwise. Typically (unless *A* and *B* are diametrically opposed), one way will be shorter than the other. Notice, however, that both these paths are curved – even the shortest route between these two points is no longer a straight line. A straight line connecting *A* and *B*, in other words a chord, can of course still be drawn. But to beings that live on the surface of the circle, this is not an allowed path; inside the circle is nothingness – they cannot tunnel through. To get from *A* to *B* in the shortest possible time, they must traverse a curved path *simply because the space in which they live* is curved!

Finally, all these separate puzzle pieces begin to drift into place. You can almost hear the sound as they snap together, each piece connecting with its neighbor, and the beginnings of a picture glimmer into view. We need to modify our conception of reality somewhat, says Einstein. It is true that objects left to their own devices tend either to stay at rest or continue their uniform motion in a straight line, but only in Euclidean, or flat, space; a more general statement is that inertial objects travel along the shortest paths they can find. In flat space, these are simply straight lines; in curved space, the analogues are called *geodesics*.

Moreover, Einstein showed – after working out a lot of mathematics – that, just as our study of accelerating systems hinted, the very shape of space changes in the presence of a gravitational field. But take a step backward – a gravitational field is set up by an object that has mass; so what we are saying in effect, is that a mass, every mass, changes the *shape* of space. Heavier masses change it more and lighter masses change it less, but every mass makes its presence felt.

This is somewhat of a leap, because we are not used to thinking of space as a tangible thing and if something is intangible, how can it curve? One can of course visualize spheres and circles and rubber sheets, but what are these made of? The answer simply is *spacetime.* Without delving into higher mathematics, apparently the most that one can say is that the arena in which we find ourselves, our entire playing field, changes shape. It is a rough analogy of course, but then we are sketching with a few words what can only be exactly stated in pages of mathematical equations.

And now it remains only to tie the loose ends up. Any massive object changes the space around it. This effect is strongest close to the object and consequently, inertial observers find themselves tracing curved paths in the vicinity of a mass. What we call gravity, said Einstein, is merely a manifestation of this warping of the geometry. If this is indeed true, if the very shape of space changes, then it is not just massive objects whose paths are affected, but light too would be forced to bend according to the dips and curves it encounters in its journey from one point to another. This is where his famous result came in, the one that erupted in headlines all across the globe. If this theory is true, said Einstein, then the light from distant stars should change its path a little when it grazes the sun. His equations even predicted the precise degree to which light should turn. When this effect was finally measured, in the solar eclipse of 1919, it was found that Einstein was vindicated.

The night I finally finished reading Einstein 's book, I walked to my little window and looked at the night sky like I never had before. Who would have suspected that starlight does not whiz nonchalantly by the Sun but in fact bends in acknowledgement? In the dark depths of space, completely unbeknownst to us, these subtle gestures have been exchanged for centuries. How many other secrets do the heavens hide, I wondered? Which others of Nature's artistic refinements are we still missing out on?

One common theme threading through Einstein 's work was unification. He took entities that were so obviously dissimilar that their distinctions were barely worth stating, and reconciled them into a larger whole. Time and space, for instance; quantities of which our perception is so drastically different, were stitched together by Einstein into a single fabric called spacetime. Mass and energy, which experience tells us are completely separate commodities, are suddenly allowed to morph into each other and the chasm between them closes. Having played so deftly with both light and gravity, and given his propensity for incredibly elegant and deep explanations it was perhaps inevitable that Einstein's thoughts turn to a 'unified field theory' that would subsume the two great forces of nature, electromagnetism and gravity, encompass them in a single explanation and bring them to a common point of origin. His dream was to come, as he wrote "nearer to the grand aim of science, which is to cover the greatest possible number of empirical facts by logical deduction from the smallest possible number of hypotheses or axioms".

During his later years, Einstein was so seduced by this goal that he spent most of his time in its pursuit, singing what he called "a lonely old song". As Physics went relentlessly in pursuit of the quantum,

Einstein increasingly fell by the wayside. The new generation of physicists, busily working out probabilities and wave functions, largely thought his work irrelevant; the idiosyncrasy of a genius whose best days were behind him. But Einstein persisted until his death. Even in his very last days, the papers at his bedside included his drafts of a unified field theory.

Why did he fail in this venture when he had been so wildly successful earlier in equally improbable realms? Mainly because he confined his efforts to bringing together electromagnetism and gravity, without venturing into the world of the quantum. Quantum Mechanics was being developed in Einstein 's life-time and he himself had contributed substantially to the subject but it was not something he was at ease with. Despite its many successes, the theory was formulated in terms of probabilities and seemed to Einstein too messy to be quite true; it did not, in his words, "bring us any closer to the secret of the Old One" – and so, he left it out.

Nature, however, did not take to this too kindly. Apparently, even a genius like Einstein was not allowed to pick and choose. He did not have the liberty of stringing together two fundamental beautiful and smooth theories without including the rebellious choppy third. It would seem that while Nature might overlook – for the moment – if you leave out what you are genuinely ignorant of, it simply won't do to turn your back upon what is known in your time.

BOHR, PAULI, HEISENBERG, SCHRÖDINGER, DIRAC: PHYSICS BY COMMITTEE – QUANTUM MECHANICS. WAVES AND PARTICLES ARE UNITED

Once again I was able to find my way to Auditorium A. I write these lines sitting in what I have come to think of as my seat, here in the middle of the third row. I am careful to close the door behind me, not because the Auditorium is off bounds, but because anyone who sees me will probably consider it a bit odd that I am studying here instead of in my quite comfortable office; and so, since I would rather stay undiscovered, my evening ritual of sneaking away to this Auditorium has become invested with a little thrill. Harmless as the activity is, my heart does beat a bit faster when I walk here, ready to modify my route at a moment's notice should I encounter people in the corridors. I always have some excuse ready to hand out to those I meet, I am either going for a walk, or a drink of water. Ridiculous as it may seem, I have begun to guard my little secret quite jealously.

When I finally reach here and close the door, there is a veritable sense of accomplishment. I settle in to my chair, which is close enough to the blackboard and yet also to the window and the photographs that line the walls. It would be difficult to say which I need more. Mostly though, I need the spirit of this room. I can sense, as if it were a tangible presence, the work that has been done here. It hangs thick in the air around me. Countless ideas have bounced off each other here and the sounds of those impacts echo still. Embossed into this blackboard are layers of equations that have been written – and rewritten - on it and looking down at me from the walls are the black and white faces of the insightful few who have in these past few years peered into the atom. How can one help but feel inspired? If one must come to grips with quantum theory, what better place can there possibly be than this room where – more than anywhere else – it lives.

If that last sentence sounds strange, I can only say that it is not quite strange enough. Quantum theory refuses to be pinned down. It will not give definite answers to any question. It refuses to say exactly how long something took, exactly how much energy was involved in a process, exactly where an object stands or exactly what the momentum of a moving body is. Everything is blurred out and made hazy. There are no certainties anymore, only probabilities, so it is only fitting that no single place be declared the birthplace of quantum theory. The best one can do is to say that the theory was born at various times in various places all over the globe. But still, if I had to calculate the coordinates of the place quantum mechanics calls home, I suspect that the wave-function would be strongest here at Auditorium A in Blegdamsvej 17.

The theory wanders these halls, spreading like a wave; it ripples through windows, diffracts through keyholes and tunnels through walls. At other times, it shoots out idea quanta in focused beams, pelting them at every conceivable surface, until the rays rebound, reflect, and finally pass into notebooks or blackboards, refracting slightly as they do so. This activity goes on always, but during the Easter conferences, the pace and intensity of the interaction becomes positively frenetic.

The last one ended two weeks ago now, but it has left behind its footprints in chalk dust all over the institute. I myself, though I was far from being a key player, have not yet recovered totally from the frenzy of those few days. For starters, there was an absolute explosion of people. They were everywhere, and never for a moment did they seem to rest. They filled the rooms, and yet spilt over; little groups of two and three were littered all over the corridors. The concept of eavesdropping ceased to exist, every conversation was public, open to any passerby. People talked on walks, across the lunch-table and even while playing ping-pong in the library!

Every year at Niels Bohr 's invitation – the world's most brilliant minds come together to re-create this strangely intense yet oddly relaxed atmosphere where they assemble quantum theory piece by piece. There is hardly ever a fixed agenda at these gatherings; discussions are organic, evolving spontaneously according to the needs of the moment. While the casual and collegial Copenhagen meetings seem far removed from the formal Solvay Councils, they germinated from the same seed: the recognition that, for the first time ever, physics had to be done by committee. In a way, these conferences are open admissions that what faces the world right now is a madness too large to be contained in a single mind.

When the first Council was announced two decades ago, it was becoming increasingly clear that the results coming in from experiments had far outstripped the explanatory power of both classical mechanics and electrodynamics. Atoms had proved themselves undeserving of the epithet bestowed upon them by the Greeks. These supposedly indivisible entities, while electrically neutral as a whole, contained equal amounts of positive and negative charge; the negative charge was distributed equally among *electrons* which ran circles around the *nucleus* – the seat of all the positive charge. The nucleus was surprisingly small, given the size of the atom. Rutherford, who "saw" it first, said it was like "a gnat in Albert Hall"!

Inevitably, a comparison was made with the heavens; the idea of each atom containing a miniature solar system had an undeniable philosophical appeal. However, it quickly became apparent that this analogy could not be very precise. According to Maxwell 's electrodynamics, a charged accelerating particle, such as an electron in orbit, would continuously emit electromagnetic radiation, thereby losing energy and spiraling closer and closer to the nucleus, eventually collapsing into it. This was clearly not the case. In flagrant violation of Maxwell's laws, atoms remained perversely stable.

Atomic spectra only added to the puzzle. It had long been known that an element, when heated, emits radiation that is as unique as a thumbprint. More specifically, if a pure sample of an element is heated until it glows and the resulting light passed through a prism, the prism acts as a frequency sieve, separating all the different colored components hidden in the light. This filtered light is not continuous like a rainbow, but instead a series of distinct bands, separated by empty space. These patterns of light, known as the *spectral lines* of the atom, had long been used to identify elements. As Bohr remarked later, the empirical laws of spectroscopy had been known for years before it was suspected that they encoded information about the structure of the atom. They were regarded simply as "the lovely patterns in the wings of butterflies", whose beauty was to be admired, but not probed. It was, however, only a matter of time before the question arose: where was this radiation emanating from, and why did each element only emit light with a particular set of discrete, separated frequencies?

This was not the only trouble light caused. Not content either with revolutionizing electrodynamics, or forcing the birth of relativity, light lured physicists into another maddening quandary. This time it was the photoelectric effect, the phenomenon whereby electrons are emitted from a metallic surface on which light was shone. The process itself was not surprising. Electrons are held in an atom by some mechanism that "binds" them to the nucleus; light supplies the electrons with energy that they use to break these shackles and escape from the atom. The problem however, was that contrary to expectation, a brighter light did not result in the production of more energetic electrons than a dim source; the intensity of the light affected only the *number* of emitted electrons, not their individual energies. In fact regardless of the intensity of the beam, no electrons were emitted at all unless the frequency of the light source was above a certain threshold. If light consisted of waves, these would keep washing over the electron; slowly building up until a large enough tide was created to wash the electron out of the atom. Clearly, the frequency of the incident light is immaterial for this mechanism; the amplitude of the wave, i.e. its intensity, would be the deciding factor. But it had been shown experimentally that this was not the case. If the electrons had resolved to wait for blue light, no amount of red light showered upon them would make them give in; the electrons were vehement in the demand for, and response toward, color. Why would this be so?

Once again, it was Einstein, with his uncanny insight into the nature of light, who answered this question. A few years earlier, Max Planck had showed that radiation emitted by black bodies[5] is quantized. Since the radiation was thought to originate from the oscillations of atoms, Planck took this to mean that the atoms themselves could not pick their energies from a continuum but instead had to choose from a discrete set of values, each of which was found to be an integer multiple of the frequency of oscillation (scaled by a constant). Extending this idea, Einstein conjectured that perhaps light too, has a similarly grainy nature. He saw a beam of light as a stream of discrete quanta; these *photons* are indivisible particle-like entities, each of which carries a unit of energy determined by the frequency (or color) of the light. An intense beam has more photons than a weaker beam of the same frequency, but the individual photons in each carry the exact same amount of energy. Electrons depend upon photons to help them escape from the atom, but the operation must be discrete. The necessary energy must come from a single photon – it cannot be built up bit by bit. The barrier erected by the atom is not something the electron can slowly and steadily climb over, it must always be covered in a single (quantum) leap.

That being so, it is easy to see why, when the frequency of the light beam is below a certain threshold, no electrons are emitted no matter how intense the beam is made; the photons simply do not have the energy needed by the electrons to break their bonds. If, however, the frequency of light is increased, the amount of energy carried by each photon goes up. When this is equal to the binding energy, an electron would absorb a photon, thereby gaining the strength to jump the barrier erected by the atom and roam free. Moreover, Einstein said, if the energy of the photon is greater than the binding energy, the electron would break loose carrying away the "excess" energy as kinetic energy of motion. This last fact was checked experimentally and found to hold true, confirming Einstein's theory.

Given the madness that came afterwards, it is perhaps a bit strange that I linger for so long, on these – comparatively sane – beginnings of quantum mechanics. But every so often, I can't help feeling that perhaps one needs to return to the root of the issue to really come to grips with it; then again, maybe that is just because from here on, there is almost nothing left to hold on to.

When you think about it, the notion of a quantum changes everything. The sweeping continuum we are used to is swept away from under our feet and replaced by a grid. Zeno's paradox becomes an oxymoron; you just can't keep covering "half the remaining distance" between yourself and something else if you live in a quantum world. Gradual transitions are no longer the norm; this is not a place in which you emerge, gradually, out from the shadows into the light – you are suddenly and unceremoniously thrown out. The seasons do not seamlessly dance into each other, night no longer melts into day; there is no infinity of in-betweens. None of these new ideas reek of the life we know. One must take a leap of imagination to get to the realm where the quantum dwells, it is a place of which we have no direct knowledge; transitions in this new world are as sudden as crossing over from life to death.

The quantum was highly unorthodox – it challenged all our previous conceptions of physics being smooth and continuous and yet, strangely enough, seemed to be borne out by experiment. This is the conundrum that caused the world's most spectacular and renowned physicists to convene at that first elite Council in Solvay. With brilliance, diligence and remarkable open mindedness, they set upon the task of deciding whether or not the quantum was to be included in the physicist's repertoire. Did Nature paint with the flowing brushstrokes of the old masters, they asked, or the tiny dots of Seurat? They were enjoined to "take so fruitful a theory seriously and subject it to careful investigations" before finally pronouncing judgment. News traveled fast; those who were present carried the message back to their own institutions and those who had no immediate link to the participants studied the proceedings of the conference. Wordlessly, the physics community united in a search to make sense of the quantum.

Niels Bohr, then a student at Manchester, was one of those immediately set ablaze. He caught the fire directly from Rutherford, who was one of the elite attendees. Rutherford had soared to fame a few years earlier for his discovery of the structure of the atom ; it was mostly empty space, he had said, with electrons whirling around a surprisingly small nucleus. His student went much deeper. The soft-spoken Dane with his vaguely embarrassed smile might have seemed an unlikely champion for the rebellious and unruly quantum theory, but he was one of its first.

We assume there is a contradiction, Bohr said, only because atoms do not obey the laws of electrodynamics as mandated by Maxwell, but why should they? Our expectations are shaped entirely by what we perceive around us, at

---

[5] Black bodies are objects that perfectly absorb all the electromagnetic radiation incident on them, reflecting none.

scales that are roughly a thousand billion times larger than atoms; we have no direct experience of that size and as such have no right to expect the same rules to carry through to sub-atomic scales.

Instead, Bohr suggested the following model: an atom has a finite number of *stationary states*, configurations "in which no emission of energy takes place, even if the particles are in motion". Each state is labeled by an integer called a *quantum number,* which can be used to determine its energy. What this meant was that in the vast spans of the atom, there are only a finite number of allowed orbits, invisible grooves, as it were, into which electrons could slip. While sliding around in a particular discrete orbit, an electron would be stable; it would emit no radiation and carry a constant energy. Radiation would be emitted only when an electron jumped from one orbit to another. The jump had to be accomplished all at once since there was no resting place in between. The difference in energy between the initial and final states (Bohr called these the *waiting places*) was carried off as radiation. Contrary to what Maxwell 's theory would dictate, the emitted energy had nothing to do with the rotational motion of the electron in orbit but instead was determined completely by the quantum numbers of the initial and final states.

This was the picture Bohr proposed, and using it, he was able to reproduce the well-known formula for the emission spectrum of Hydrogen. Bohr's model of the atom had remarkable success and soon came to be widely accepted. Before long, however, it was found that there was some discrepancy between this theory and the more sensitive experiments. At this point, Arnold Sommerfeld came to the rescue, playing Kepler to Bohr's Copernicus[6], suggesting that perhaps the orbits of the electrons were elliptical rather than circular. Whereas a circle can be completely specified by its radius alone, three parameters are needed to pin down an ellipse; consequently, states would be labeled by three quantum numbers. The theory, thus modified, worked so beautifully that Sommerfeld was moved to compare the spectra to "a veritable atomic music of the spheres, a carillon of perfect whole number relations, an increasing order and harmony in multiplicity".

By this time, quantum mechanics itself was beginning to sound like a complex piece of music, sung in many parts. While this added vastly to the richness of the sound, the voices were not always in harmony; in fact some notes were decidedly discordant. But, looking back, I cannot see how quantum theory could possibly have progressed as far as it has, in this short a time, had the melody not been tossed back and forth so many times between such different voices.

One of these new voices belonged to Wolfgang Pauli. Pauli was first drawn to quantum theory as a student listening to Sommerfeld 's lectures in Munich. Even then, he was not seduced by the "number mysticism" everyone else advocated; in his opinion, they were all groping about in a mist and Pauli was not one to hold back on expressing his opinion, or to mince his words while doing so. However, since he was extremely incisive and his criticism often cut right through to the heart of the issue, his sharp, sometimes scathing, comments soon came to be treasured and in the small community of quantum mechanics, ideas were not considered developed until they had passed Pauli's test[7].

For almost a decade following Bohr 's initial papers, most of the work done centered on the mysterious quantum numbers. Sommerfeld had brought the count up to three, but it turned out that even this was not quite enough to explain the spectra of elements other than Hydrogen. Eventually, Pauli arrived at the number four. The problem with the fourth quantum number was that, unlike the earlier three, it seemed to defy interpretation. It did not correspond obviously to any known physical quantity and moreover, it was a *half-integer.*

This delighted Werner Heisenberg, who was a friend of Pauli 's and a fellow student of Sommerfeld 's. A few years earlier, when Heisenberg suggested the existence of half-integer quantum numbers to explain a phenomenon known as the *anomalous Zeeman* effect, Pauli had criticized the proposal in his characteristic manner saying, "Now you introduce half quantum numbers, then you will introduce quarters and eights as well until finally the whole quantum theory will crumble to dust in your capable hands". When a twist of fate had Pauli postulating a fourth quantum number which was a half-integer, Heisenberg could not contain his glee. He wrote to Pauli saying how pleased he was that "you have raised the swindle to unimagined, dizzying heights (by introducing single electrons with four degrees of freedom) breaking in the process all previous records you accused me of holding".

---

[6] *Editor's note*: See the article "*Does History Repeat Itself in a Non-Cyclic Universe?"* for Kepler's story.

[7] Famously, Pauli once said of a paper that was presented to him: *Das ist nicht nur nicht richtig, es ist nicht einmal falsch!* "Not only is it not right, it's not even wrong!"

The teasing aside, it appeared Pauli really had stumbled on something fundamental. This something, the so-called *Exclusion Principle*, said that every atom has a discrete number of states, labeled by a unique set of four numbers; since each state has room for only one electron, any two electrons in an atom will be in different states and hence will carry a distinct set of quantum numbers. Electrons see these states as being arranged in a ladder-like structure, with higher rungs corresponding to larger energies; any given electron will tend to minimize its energy by occupying the lowest rung available to it. Since the quantum numbers belong to the state (the rung) not the electron that occupies it, electrons can change their quantum numbers by jumping to a different state.

This simple law explained the distribution of electrons in an atom and indeed – as was later discovered – in any system. It is almost as if electrons, when placed in a group, distinguish between each other by putting on unique badges bearing their quantum numbers. The problem of interpreting these numbers remained. Three of the quantum numbers had a simple correlation to the elliptical orbital motion of the electron in three dimensional space. The fourth was something that could not be made to correspond to any known movement until George Uhlenbeck and Samuel Goudsmit, two young students of another of quantum mechanics ' greats; Paul Ehrenfest, proposed the concept of *spin*. Drawing once again on the solar system analogy, they conjectured that, like the Earth, the electron also spun on its own axis and the two values (+½ and -½) of this quantum number corresponded to the electron rotating clockwise or anti-clockwise. Their proposal met with a remarkable amount of success and Pauli is said to have regretted not putting this explanation forward himself. It had occurred to him too, but he had brushed it away; apparently his overly critical nature sometimes got in his own way.

The more I reflect on the amazingly jagged route in which it progressed, the more firmly I am convinced that Quantum Mechanics is a song that just had to be sung in many voices. There was a part for passionate mathematicians and a part for those who thought in pictures; there was a part for philosophers and one for those who loved formalism; a part for the extraordinarily intuitive and one for those who had a rigorous bent; a part for those with many ideas and one for those who could criticize them; a part for those who poured over experimental results and one also for those who focused on theory alone. I look up to my right, at the wall full of pictures and see the faces of this great chorus staring down at me. Each distinct, intense face is etched in my mind from the many times have I stared at these static pictures and the many times have I actually seen the people themselves, in all their animated zeal. But right now, in the fading light and from a distance, all I can make out is a haze of black and white. From somewhere in this mist where individual features are lost and boundaries blur, a faint hum emanates and it hits me afresh that the song of quantum mechanics had to be sung thus. No single man, no matter how talented, could have sustained all these distinct voices in his head without going mad.

This, perhaps, was where Bohr 's greatest contribution came in. His "University Institute of Theoretical Physics" became the place where all these different voices came together and sang their parts. Melodies were modified, battles occurred not just over discordant notes but also chords and scales. Both for people who wanted an audience for their songs and those who merely wanted to be in the presence of echoes and resonance, Bohr's Institute was a haven. After his 1922 lectures on atomic theory at Gottingen, at what came to be known as the *Bohr Fest*, Bohr's popularity soared and he was able to attract, on the basis both of his work and also his remarkable humanity, a host of the most talented minds to Copenhagen. Pauli was one of those who were drawn instantly to this hive of activity; Heisenberg was another. The atmosphere at the Institute was scintillating but relaxed. Bohr's deeply creative spirit set the tone for collaboration since he, himself, was incapable of thinking without human company. All his work was sculpted in conversation. Ideas were tossed back and forth ceaselessly at Blegdamsvej, keeping pace with the ping pong balls in the library. Bohr's incessant discussions kept the momentum going while Pauli, with his genius for criticism served as the referee, keeping score and calling fowls.

Pauli fast became Bohr 's favorite sparring partner, and Heisenberg a cherished friend, but the *modus operandi* of his two young "disciples" differed widely from that of Bohr. As Heisenberg later recalled "Bohr's insight ... was not the result of a mathematical analysis of the basic assumptions, but rather of an intense occupation with the actual phenomena, such that it was possible for him to sense the relationship intuitively rather than derive it formally". In other words, Bohr was primarily a philosopher; Heisenberg and Pauli were not. By 1925, these two were convinced that a radical reformulation of quantum mechanics was in order, and I am not sure that Bohr entirely disagreed. He fully recognized the limitations of language and the inherent inability of our vocabulary to encompass the atom. Since our words are shaped to describe classical concepts, we cannot hope to explain anything precisely but must, like poets, be content to "create images and establish connections".

Imprecise word pictures were simply not enough to satisfy Heisenberg or Pauli. Merely tweaking things here and there did not suffice; the theory had to be re-cast entirely. Pauli felt that the atom must not be placed "in the shackles of our prejudices (of which in my opinion the assumption of electron orbits … is an example)". Heisenberg went a step further, saying "it seems sensible to discard all hopes of observing hitherto unobservable quantities, such as the position and period of the electron, and to concede that the partial agreement of the quantum rules with experience is more or less fortuitous". Together, they argued for a formalism that dispensed with the cosmic analogy, and in fact with any pictorial representation. Almost the only thing that to which they did not take exception was the fundamentally discrete nature of the quantum.

What Heisenberg suggested was not a model of the atom but a completely new philosophy of physics. He proposed that theories should avoid mentioning quantities that were not experimentally accessible and be formulated strictly in terms of observable quantities. In essence, Heisenberg said, we should control the urge to "connect the dots" and paint a picture of what goes on behind the scenes, or to construct a mythology for why things are so. We should accept that we will never know for certain what goes on in the hidden realms "backstage", and so we should limit our reflections and concerns only to what we can directly experience.

In that spirit, Heisenberg stopped treating electron orbits as something that could be visualized and replaced the picture by a mathematical equation that was an appropriate abstraction of Newton 's laws. He restricted himself to considering only physical variables, for instance positions and velocities of electrons. The challenge was to see if this new formalism encapsulated enough information to be able to yield the atomic spectra. In order to do this, Heisenberg needed to find a way of describing the radiation emitted by electrons as they transitioned to different states.

Even earlier, in Bohr 's more mechanized model, no one had been able to predict just when a particular electron would decide to leap between orbits. Statistical laws had been formulated, however, and these worked surprisingly well. Even in the absence of any claim regarding the will of a particular atom, one could predict with stunning accuracy the behavior of a crowd. Mob mentality is amazingly predictable, even at the quantum level. Heisenberg worked this uncertainty into his theory by introducing what he called *transition functions* which measured the probability that a given electron would "transition" from one specific state to another. The logic was that transitions which had a higher probability would happen more often and the corresponding lines in the atomic spectrum would be more intense.

There were no pictures to guide him anymore, all that was left was mathematics and so Heisenberg followed his equations. He had not progressed very far before he realized, much to his dismay, that the familiar rules of algebra did not seem to hold anymore. The quantities he was working with displayed a strange property; when multiplied in one order, they yielded a different answer than when multiplied in another; the *order* in which the operation was carried out, was paramount. This completely violates our experience of mathematics and even Heisenberg, with his insistence on resisting the temptation to fill in gaps with preconceptions, found this idea almost too much to swallow; perhaps it signaled a fundamental problem with the theory, he thought. Pauli, however, gave this work his blessing, going so far as to call it the *morgenrote* – the beginning of the dawn – of quantum theory. As a further vote of confidence, Pauli used Heisenberg's formulation to work out the energy spectrum of the hydrogen atom.

As word spread in the community, other people and their opinions came to bear on Heisenberg 's work. This, so-called, *noncommutative* nature of multiplication did not quite strike everyone with the same force as it had Heisenberg[8]. In particular, Max Born at Gottingen realized that these "new" multiplication laws causing Heisenberg such dismay, had actually been long known to mathematicians and were in fact the necessary prescription for multiplying the array-like objects known as *matrices*. While Born worked on this theory, together with his student Jordan and Heisenberg, far away in Cambridge another of this era's legends stepped up to take his place on stage.

Paul Dirac did not have one of those loud, larger than life personalities that would automatically attract attention, but when he joined in the quantum mechanics chorus, the notes he struck were so pure and so sure that they cut through the hum, and all around him stopped to listen. The song he sung was almost a hymn to mathematics; Dirac had little time for literature or the theater, philosophy he dismissed as being a waste of time, but he was nevertheless fascinated

[8] *Editor's note*: See footnote on page 67.

with the nature of beauty and nowhere else did he find beauty as perfect and majestic as in mathematics. Unlike his contemporaries, Dirac worked in solitude. Perhaps it is only words that need company, only philosophy that needs conversation and interpretation. Mathematics is content with itself. It is capable of internal checks, and has no need for external approval. He framed his questions not as philosophical enquiries into the nature of phenomena, but instead as equations which could only have a single precise meaning. When something is shown to be mathematically true, it leaves no room for emotion, or dissention or opinion; a mathematical truth is absolute and timeless.

Dirac approached Heisenberg 's work with a perspective that was uniquely his own. Quickly, he decided that the crucial feature of this new theory was not its construction in terms of experimentally observable quantities, but instead the statement it made about the non-commutation of physical variables. And so, he says "I was led to concentrate on the idea of non-commutation and to see how the ordinary dynamics which people had been using till then should be modified to include it". With this in mind, Dirac turned to Bohr and Sommerfeld 's theory of atomic orbits and found that its mechanics could be expressed in a "classical" way that had been discovered almost a century ago by a scientist named Hamilton. The virtue of this formalism was that it could rather straight forwardly be generalized to non-commuting variables. Dirac exploited this to re-express Heisenberg's matrix mechanics in a familiar "classical-looking" form, with the crucial provision that physical variables no longer commute. The Gottingen group took a different approach but they too, arrived at the same results. Perhaps the reason Dirac's formulation prevails, is that it bridged the gap between Heisenberg's matrix mechanics and the vastly different, but equally effective, wave equation put forth by Schrödinger.

For a while though, before Dirac showed that these formulations were equivalent, there were two candidates competing to be *the* theory of quantum mechanics ; the confusion over methods and interpretations had burst into a full blown battle. In 1926, an Austrian named Erwin Schrödinger sent tremors throughout the physics community with his *wave equation*. Schrödinger drew on an idea that had been set forth a few years earlier by a French nobleman called Louis de Broglie. The idea itself was simple enough, but its ramifications were enormous and far reaching. de Broglie by-passed the entire discussion on the atomic model and went back to the beginnings of quantum theory and the existence of light quanta.

The wave nature of light had been well established for centuries; light undeniably participates in phenomena such as diffraction and interference, both of which are behaviors that cannot be emulated by particles. And yet, Einstein said, not all of light's tricks can be explained by appealing to its wave nature alone; at times, light simply must be interpreted as a stream of particles. de Broglie turned this around and asked why, if a wave can exhibit a particle like behavior, particles can then not also exhibit wave-like properties? While the question itself might seem to be perfectly logical, it is so at odds with experience that no one thought to ask it earlier. Even after quantum mechanics had been shattering preconceptions at almost every step for two decades, there was apparently still an inertia in the mind … seemingly "obvious" questions still remained unasked.

Luckily, this question, once asked, did not take too long to be answered. The mathematics involved was easily worked out and de Broglie found that indeed, a wave could be *associated* with every material object. The reason this had not been discovered earlier, he said, is that objects at everyday scales, have exceedingly tiny wavelengths and consequently their wave nature is somewhat suppressed. Smaller objects like electrons, on the other hand, have long wavelengths and hence openly exhibit wave behavior.

Using this construction, de Broglie was able to arrive at a charmingly simple "derivation" of the discrete energy levels of electrons in an atom. Drawing an analogy with stretched strings, de Broglie said that just as a string of a certain length can sustain only a particular set of notes, the same is true of electrons in orbit; stable states correspond to standing waves. In tracing an orbit, the electron-wave must go through a number of complete cycles so that it seamlessly joins back onto itself; anything else would lead to the wave interfering destructively with its own self, and hence decaying. These *waiting places* of the electron turned out to be far from arbitrary; they are simply orbits whose "length" is an integral multiple of the wavelength of the electron-wave.

But even though the equations said that matter and waves were two different sides of the same coin, this conclusion was extremely problematic. After all, if matter could be *smeared out* into waves, what were these waves made of and how were they to be interpreted? How is it that a single object can exhibit both particle and wave behavior when we know full well that particles and waves are mutually exclusive categories?

There are so many parallel threads interweaving here, so many diverse concepts coming together, such a lot of work being done by such vastly different people in various scattered places … so much simultaneous activity, that I hardly know which idea to think through first; the proper flavor of quantum mechanics can only be conveyed by multiple voices talking at once but the noise is overwhelming and I find myself being grateful for the linearity of speech which forces just one word out at a time, and hence enforces order on my tangled thoughts.

Around the time Schrödinger was working on his wave equation, Heisenberg also came up with his masterpiece, the one statement that ensured him a place in posterity. And when immortality came, it was, surprisingly enough, cloaked in the annoying guise of non-commuting variables. Heisenberg discovered that when a pair of variables does not commute, the values of both cannot be exactly determined simultaneously; either we must content ourselves with a small uncertainty in the measurement of each, or else we can pin down one of these quantities exactly, while leaving the other undetermined. Position and momentum are an example of such a pair – the more precisely we know the position of an object, the less certain we are about its momentum. This came to be known as *Heisenberg's Uncertainty Principle*. As some paraphrased, this implies that the very act of measuring its position disturbs an object such that its momentum is no longer certain.

Even though the uncertainty is rather small, the philosophical ramifications of this statement are huge. What it really says is that one can never know a state exactly or perhaps that a state with both a definite position and a definite momentum simply does not exist. This is completely different from the romantic longings of earlier years. While one had never really specified all the parameters of a physical system with perfect precision, the assumption remained that this was possible in principle. The uncertainty principle says that that is no longer true; it defined an edge to what was knowable, even in theory.

Understandably enough, people had a lot of trouble swallowing this statement. Pauli wrote to Heisenberg in exasperation asking him to help make sense of the fact that "one may view the world with a p-eye and one may view it with a q-eye, but if one opens both eyes at the same time one goes crazy"[9]. Effectively, the uncertainty principle signaled the end of determinism. As Dirac put it, "In the statement 'if we know the present, then we can predict the future', it is not the consequence but the premise that is false. As a matter of principle, we cannot know all the determining elements of the present".

In response to this crisis, quantum mechanics ' philosopher came up with his theory of *complementarity*. Bohr said, "It is wrong to think that the task of physics is to find out how nature is. Physics concerns what we can *say* about nature". Perhaps the contradiction is not inherent in nature, but instead in the way we have defined certain concepts. To us, waves and particles are fundamentally different constructs, but apparently an object can have both these attributes at the same time; depending on how we probe the object, we will be shown either its wave face or its particle face but never both at the same time. It is almost as if, when we ask nature a question, we are looking through a filter – what we see depends on the color of the filter we impose and is only a part of the whole picture. The "truth" is wider than we can encompass in a single description; the whole is more than the sum of its parts and Nature seems to have drawn a distinct boundary around what we can know.

But back to Schrödinger and, once again, the epic confrontation between images and equations. It seems as if the two go hand in hand, like waves and particles; perhaps mathematicians and visionaries see different aspects of nature, and it is only by piecing their complementary images together that the puzzle can be completed.

Heisenberg 's matrix formulation of quantum mechanics was something Schrödinger rebelled against, claiming to be "disgusted, if not repelled, by what appeared to me to be a rather difficult method of transcendental algebra, defying any visualization". The dual nature of matter, as unsheathed by de Broglie, appealed strongly to Schrödinger. Since objects at the atomic scale are small, their wavelengths are large enough to be appreciable and so, Schrödinger argued, quantum mechanics should properly be formulated as a theory of waves. To lend support to this view, he drew an analogy with optics. Much of our theory of light ignores its wave nature, and one can understand reflection and refraction and images formed in lenses perfectly well, by simply using ray optics. But light also exhibits phenomena like diffraction and interference, and to understand these we need to invoke the wave description.

[9] The mathematical symbols for momentum and position are p and q respectively.

Perhaps, said Schrödinger, mechanics also works analogously. Perhaps there is ordinary mechanics, like there is ray optics, and it can be used to understand classical phenomena, but it is not yet the whole answer; for certain other phenomena, the only possible explanation stems from an analogue to wave optics, what he called *wave mechanics*.

The inevitable question, of course, was that if these waves exist, how are their dynamics governed? Just like Newton 's laws of motion dictated the trajectories particles can trace, a wave equation would be needed to trace out the paths of waves. This is precisely what Schrödinger found; he wrote down the dynamical equation of quantum mechanics, a deceptively simple symbolic relationship that encodes the evolution of a quantum system. The solutions of this equation, each of which had a definite energy, exhibited wavelike properties and came to be known as *wave functions*.

All this was wonderful, an absolute triumph in fact, but even such a beautiful equation could not cloak quantum mechanics ' contrary nature. Definite answers were not forthcoming here either, and once again, Nature revealed only statistical averages; probabilities could, in fact, be read off simply from the shape of the wave-function. One could, for instance, set up Schrödinger 's equation for an electron in an atom. The solutions of this equation, however, would not be definite trajectories *à la* Bohr ; rather than being able to trace with certainty the path an electron would follow, the best we could do is predict the *likelihood* of the electron being at a particular place at a particular time.

It was almost as if the orbits of old, inked onto the page in sure defined strokes, had now been smudged and smeared. In the cathedral of the atom, the electron was no longer a fly but instead a vague surrounding buzz, stronger in places where the "fly" was more likely to have been.

Suddenly it appeared as if the landscape of reality was painted in watercolor. Paints bleeding, diffusing, melting into each other, thinning out to where it is almost impossible to say where something ends. It was disconcerting to say the least. Everyone struggled to make sense of this unexpected universe we had unknowingly inhabited for so long.

One of the best expressions of this intellectual frustration came from Cambridge, in the form of an anonymous poem some students tacked on to the walls of Cavendish lab. It was a "petition", made by electrons who were desperate to be set free from the "dread uncertainty" of the "hated quantum view". They went on to lament:

Once in a pleasant order *our smooth-flowing time was spent*
As the classical equations *told us where to go, we went*
We vibrated in the atom, *and a beam of light was freed*
And we hadn't any structure – *only mass and charge and speed.*
We know not if we're particles, *or a jelly sort of phi*[10]
Or waves, or if we're real at all, *or where we are or why*

The electrons' nostalgia for the classical deterministic past was shared by many. But more and more people were becoming resolved to the fact that on quantum scales, nature behaves so differently that our everyday experiences and intuition simply cannot form the basis for a satisfactory model. Schrödinger went so far as to say that such a model was not even thinkable. "Or, to be more precise, we can, of course, think it, but however we think it, it is wrong; not perhaps quite as meaningless as a 'triangular circle' but much more so than a 'winged lion'!"

I lift my gaze again, towards the photographs on the wall. What a Pandora's Box these men have opened! They have chopped up nature into miniscule pieces, each of which must be treated as an indivisible whole. They have drawn boxes around everything; invisibly small boxes, of course, but the fact remains that everything is separated, contained. There is no process anymore, no stages of "becoming"; there is only being – and nothingness. They do not allow continuity anymore, there is no seamless flow left; hidden within every apparently smooth curve are tiny straight lines which turn corners, there is a pointed, angular awkwardness to it all; you feel its stinging graininess as a strange surprise, almost as if you found sand under your feet when you expected marble. They have completely

[10] The Greek letter phi (Φ) was the symbol used to represent the wave function in Schrödinger's equation.

smudged concepts that were perfectly clear earlier. Infinite shades of grey have been introduced into pictures that were beautifully defined in black and white.

Looking up makes me realize how dark the room is now. Twilight is melting slowly into night and I can no longer make out the equations on the blackboard, but even now I can feel their pull. How potent they are, these scribbled symbols, these dim one-dimensional projections of a multi-faceted Technicolor reality! How intensely they draw you in, how much discipline and loyalty they demand … ridiculous that one would need to pledge such unwavering allegiance to chalk dust … and yet the rewards they bestow to the faithful are so tempting, so full of wonder, that one cannot help but be seduced by these equations.

Time and time again, it has been shown that, if you are loyal, and you follow the circuitous path they draw you into, if you watch them perform their intricate gymnastics patiently, slowly their ways become clearer. They speak in a language all their own, but it is obvious that, always and forever, the equations know what they are saying. And so, I sit with them in the dark another few minutes, letting their wave-functions intermingle with mine.

THE STANDARD MODEL: TRIUMPH OF GAUGE THEORY; ELECTROMAGNETISM MEETS THE WEAK FORCE

Trieste,

Summer 1981

My darling children,

I knew I would miss you even before I boarded the plane, but I did not realize how much I would ache to show you and your mother the beauty I now behold. The first day I was here, I found myself turning around every few minutes to look for you … If God wills, one day the four of us will visit this wondrous city together. For now, let my words be your eyes and ears, and let us share experiences thus.

It is the lunch hour, and there is a break between talks. Just for today, I have skipped the long queues in the cafeteria and crept away to Miramare Park, so that I can sneak in a little conversation with you. The park shares a wall with the ICTP and one can go from the lecture hall to a stone bench in five minutes. This is such a breathtakingly beautiful place to come and think in; a fairy-tale castle hovers over the brilliant blue Adriatic Sea and the park itself is full of winding paths and little secret places in which to sit while the ecstatic new ideas in my head cease their clamor and slowly, begin to settle down. In this place of infinite calm, I understand what Bohr meant when he said that "part of infinity seems to lie within the grasp of those who look across the sea". A deep peace permeates me, rejuvenates me for the intellectual onslaught next door.

It is not only the seminars that are exciting - just being in the presence of so many people who are interested in the same things is energizing. I had not quite realized research could be like this, but Abdus Salam says it can be no other way. "Theoretical Physics is a subject in which speech is the important thing, not the written word" he says. "You have got to go around and talk with people and be in contact if only to learn that this particular mess of papers here on my desk is rubbish and these others are the important thing. You can scan the whole blasted lot and never find out what is important but going to an active place for a day, you can easily get to know the significant from the insignificant. So a man living in isolation or with a small group has every chance of just deteriorating." It was this belief, grounded in his own experience, which led him to create the International Center for Theoretical Physics ; a little haven where scientists from the Third World come together with their colleagues from developed countries, for Salam firmly holds that "scientific thought is the common heritage of mankind".

I am trying to talk to as many people as I can, to work through my problems, to make up the discrepancies in my knowledge and actually, even to identify gaps that I did not realize existed. It is amazing how helpful people are, and how generous they are with what they know. A group of us get together in the evenings to talk about what we heard in the seminars. Everyone brings their unique strengths and varied backgrounds to the table and we each emerge from our discussions so much the richer. Now that I have experienced it, I find that this air of collegial collaboration is integral to progress and I cannot help but feel that this is how science should be done. And just as it is becoming clear to me how one should work, it is also becoming clear what I want to work on. In all the talks I have heard so far, what fascinates me the most is the *Standard Model*, a theory of particle physics that draws into its welcoming folds the quantum constituents of our universe and the forces that act between them.

It is a source of endless fascination to me, this human compulsion to *organize* all we see, to sort things into groups and to bring order to apparent chaos by finding something fundamental which, through various combinations and contortions, gives rise to the wide and varied world we perceive; we are forever tracing the tree back to the seed. Few things are more intellectually or aesthetically pleasing than the ability to collect seemingly diverse phenomena and unite them through a single explanation. We enjoy and appreciate the richness of structures even more if we are aware that their complexity stems from a fundamental underlying principle; we seem instinctively to want economy – not in structure, but in principle.

Feynman, the larger than life, bongo-drum playing, safe-cracking physicist who[11] wrote down the quantum theory of electrodynamics, put it quite eloquently: "As long as it looks like the way things are built is wheels within wheels, then you are looking for the innermost wheel - but it might not be that way, in which case you are looking for whatever the hell it is that you find!" In fact, physicists are even more ambitious – they are looking not only for the wheels, but also asking *what* makes them turn; for when you break the world down to its constituents, these two basic categories are what remain – matter and the forces that act upon it.

It appeared, at the beginning of this century, that the puzzle had been solved. Atoms were found to consist of tiny negatively charged particles called electrons "orbiting" a nucleus made up of positively charged particles called protons and objects with no electric charge, called neutrons. Electrons, neutrons and protons were considered indivisible and hence thought to be the building blocks of atoms and in turn, of all matter. The two fundamental forces of gravity and electromagnetism sufficed to explain every phenomenon that had been explained thus far and for a while, it seemed as if these were the only two forces that existed. The idea of somehow unifying these into one basic force tantalized Einstein and he spent his last few decades pursuing this goal.

Seductive as the idea was, it had its detractors. Pauli, one of Quantum Mechanics' geniuses, brushed it aside, saying "what God has cast asunder, let no man put together". Even though Einstein 's search remained unsuccessful, the seed for a *Unified Theory* was planted in the collective mind of the physics community and ever since then, many ideas have sprouted, each claiming to be *The One*; as Freeman Dyson said, "The ground of physics is littered with the corpses of unified theories". But until the Standard Model, nothing stuck.

By the time the Standard Model of Particle Physics was formulated, our understanding of both matter and forces had undergone a quantum leap. By probing the atom, we had realized that our previous knowledge was incomplete. While gravity and electromagnetism were enough to justify macroscopic phenomena like the formation of stars and galaxies and the structure of the elements, they could not explain what went on inside the atom. Drawing on these forces alone, there was no reason for the nucleus to stay together; the electrostatic repulsion between protons would cause them to fly apart, and no atoms other than hydrogen could ever be stable. Another phenomenon that could not be encompassed by the two-force paradigm was the radioactive decay of a neutron into a proton and electron[12]. Clearly, at the nuclear range, new forces were coming into play.

It was this line of questioning that eventually led to the discovery of two new forces and a deeper layer of substructure hidden within the neutron and proton. Of the original trio of building blocks, electrons alone continued to be *elementary* in that they still seemed indivisible. In our revised post-quantum understanding of the world, protons and neutrons were *compounds* of elementary particles given the curious name *quarks* and the *strong force* was the glue that bound quarks together, keeping them contained within the confines of a neutron or proton. Radioactivity appeared to be the process whereby one type of quark changed into another and the *weak force* was called upon to govern this phenomenon. But the new force alone did not suffice; a satisfactory explanation of radioactivity demanded the existence of an entirely new particle. Based entirely on theoretical considerations, Pauli postulated that this hitherto invisible particle – the *neutrino* – must exist, and in a historic testimony to the power of theory, this was soon found to be true!

---

[11] Together with Schwinger and Tomonaga.

[12] The source of the sun's energy was yet another phenomenon that could not be explained by only gravity and electromagnetism.

With these amendments to our model of the universe, we could once again explain everything in the realm of our experience – whether directly, through our senses, or indirectly, through instruments like telescopes and particle detectors.

In this revised picture, matter was divided into two classes – quarks, that felt the strong nuclear force and leptons (like the electron and neutrino), that didn't. As far as the fundamental forces were concerned, their tally had risen to four. This fact was so surprisingly well-known that the news had filtered through even to the modest village school in Pakistan where Abdus Salam spent his early years. In an amusing anecdote, Salam recalls the first time he heard of the nuclear forces. "When I was in school in about 1936, I remember the teacher giving us a lecture on the basic forces in nature. He began with gravity. Of course we had all heard of gravity and Newton 's name had permeated even to a place like Jhang. Our teacher then went on to talk about magnetism and showed us a magnet. Then he said, 'Electricity. Now there is a force called electricity but it doesn't live... [here] ...It lives in the capital town of Lahore, a hundred miles to the east. He had just heard of the nuclear force, and he said 'That only exists in Europe'. Since it did not exist in India or Pakistan, we were told not to worry about it."

But worry about it he did. In fact, a few decades later, it was Abdus Salam – along with two other scientists, Sheldon Glashow and Steven Weinberg – who took the first major step in making Einstein 's forgotten dream a reality. What these three people were able to show, independently of each other, was that electromagnetism and the weak force had, in fact, started out as the same force when the universe was young, the so-called *electroweak force*. As time passed, these forces developed somewhat asymmetrically – much like identical twins who look indistinguishable at birth, but develop their own unique personalities and appearance as they grow up to the point where they barely appear to be related! Since our experience of these forces is as they are now and not as they used to be, it appeared to us that they were inherently different whereas they can be traced back to the same root. As the physicist writer John Barrow says, "We do not observe the laws of nature, we observe their outcomes. Since these laws find their most efficient representation as mathematical equations, we might say that we see only the solutions of these equations, not the equations themselves".

The coming together of fundamentally different entities has always led to a leap in understanding; this is only inevitable, since in learning to view two previously distinct objects as part of the same framework, we are necessarily breaking down some mental barriers that cause us to perceive the two objects in separate ways. What made electroweak unification possible was the fact that both nuclear forces, and electromagnetism before them, were described by *gauge theories*. The beauty of gauge theory is that it interweaves particles with the forces they experience. Earlier, these were distinct problems with particles and forces being treated separately. In gauge theory, the two emerge together – for any given force, we are *told* by the theory what kinds of particles it can affect. Just as electric charge measures the ability of a particle to participate in the electromagnetic interaction, physicists use the words *color* and *flavor* to measure the abilities of particles to feel – and respond to – the strong and weak nuclear forces. Another fascinating feature of gauge theories is that forces are mediated by particles (called *bosons*) that travel back and forth between the matter particles. The messenger of electromagnetism is the *photon*, the weak force is carried by the W^{+}, W^{-} and Z bosons, whereas the carriers of the strong force are known as *gluons*.

Essentially, what Glashow, Salam and Weinberg showed was that at the high energies which were present in the early universe, there was perfect symmetry and the W and Z bosons were indistinguishable from the photon. As time passed and the universe expanded and cooled, this symmetry was broken. Mathematically, I am told, the theory is quite lovely, but with my current grasp of the subject, I can not quite see this for myself. One of the most important lessons I am learning here this summer is the fact that, especially at this level of abstraction, physics confides only in those who speak mathematics; if, like me, you speak it haltingly, you get the basic meaning but miss out on the subtleties.

At times, during the seminars, I feel as I do when I am out shopping in town. Trieste is a small place and most people here speak only Italian, so despite the language guide I keep referring to, I often find myself making elaborate gestures to get my point across. When my meaning is finally conveyed, the townspeople – who are remarkably friendly and patient – direct at me a barrage of fluent flowing Italian. I understand perhaps every twentieth word of what they say, but I nod and smile nonetheless, for the feeling of having made a connection is quite wonderful. My feeling during the more technical seminars is similar; as far as I am concerned, the lecturers might as well be speaking a foreign language. I understand just enough, every so often, to get a vague idea of what they are saying, but what they talk about is so beautiful, so wondrous, that it keeps me hooked and I am resolved to learn the mathematics so that I can speak it like the "locals", and participate in these subtle nuanced conversations. For the moment, while I attempt to work out the mathematics for myself, I am also making a concerted effort to focus on the bigger picture, so that I do not lose myself in the lines between calculations and forget why I started them in the first place. To this end, I often treat myself to short breaks, browsing the aisles in the library, randomly taking off the shelves any title that speaks to me. In a book[13] I came across yesterday, there was a passage I found particularly poignant: "The universe", it read "began in a state of absolute symmetry in which there was neither place nor time nor varieties of particles or forces. A cosmic fall from grace came during the first fraction of a second of the expansion of the infant universe when the symmetries of genesis fractured as the universe cooled ... At higher energies, symmetries are manifest, while at low energies they are hidden. This fall from grace is essential to existence as we know it. Absolute symmetry is beautiful but sterile. Every motion breaks the symmetry of space, creating a here different from there. Every event breaks the symmetry of time by creating a difference between before and after. One must break symmetries in order to get things done".

The unification of the electromagnetic and weak forces was a phenomenal achievement and it resulted in Glashow, Salam and Weinberg being awarded the Nobel Prize in 1979. The Nobel Prize is more than an award for excellence – it is a nod of acceptance by the physics community and signals the inclusion of the theory into the standard core of knowledge. As such, before the Prize was awarded, the electroweak theory had been tested rigorously and had proved to hold. In my opinion, however, there is another argument for its truth and importance. The fact that Glashow, Salam and Weinberg, each with their unique personalities, beliefs and motivations had – working independently of each other – "stumbled" upon the same theory.

Salam 's motivations were almost religious. "My whole background from Pakistan was that of mathematics and theory" he says, "and I think the emphasis on symmetry is something which I have inherited from the culture of Islam. The belief in unity, in there being one simple cause for all the forces that we see, has a basis in my spiritual background". Weinberg, on the other hand, considers religion 'an insult to human dignity'. "The more the universe seems comprehensible, the more it seems pointless" he says. If two people who are so different to each other can come up with the same theory, surely it cannot be a subjective construction of someone's mind, but rather, a statement of some objective reality that exists *out there*. Scientists are more like archeologists than artists, not painting pictures whose meaning they alone can see, but excavating fundamental truths, long hidden.

Once electroweak unification was achieved, it was of course irresistible to ask whether this process could be repeated again, such that all four forces were eventually unified into one! While this has not yet been accomplished, some Grand Unified Theories have been proposed that unite the strong force with the electroweak. In these scenarios, the universe starts out by treating three forces on an equal footing and it is only with time – and the breaking of the perfect symmetry – that they diverge. Gravity, however, refuses to give in to this scheme. While the three forces can be described by quantum field theories, gravity resists all efforts to be *quantized*. For the purposes of particle physics, this does not matter much since the objects we are dealing with are so incredibly tiny that the gravitational attraction between them is negligible anyway. However, intellectually, of course this poses a problem and is a possible obstacle to a true *Theory of Everything*.

Will a unification of all four forces ever be achieved? Something within me says *yes*, but I know I need to understand the problem more precisely before I am entitled to voice any opinion at all. I want to work through the mechanics of the Standard Model and understand the calculations of quantum field theory myself. Particle physics, I know, is what I want to spend my time on.

Like a camel storing up water for the long dry walk ahead, I am thirstily trying to soak up all the knowledge around me; during the more technical talks, I am afraid all I really soak up is the atmosphere. But I am bringing all the relevant material home and am resolved to work through it so that if I get the chance to return next year, I will come back better prepared. Difficult as it might be to master the background, I know that if I can hold on to the excitement I feel at this moment, I will sail through; and so, I am surrounding myself with inspiration, fanning the flame

[13] This excerpt is from Timothy Ferris's essay "Next – the coming era in Science".

vigorously, trying to ensure that it will at least flicker for an entire year until I can come back and ignite it again. Hopefully, the three of you will be with me then, to experience the beautiful majesty of both this place, and the noble pursuit people engage in here.

I have always held this to be true, but here in Trieste I have seen my theory borne out – science is the great equalizer. In this province of abstract, conceptual thought, all social inequalities are left behind. All that matters is the merit of one's ideas. This is one place where those of us from developing countries can, if we try, stand shoulder to shoulder with our colleagues from the developed world. If we work hard enough, it is completely within our power to push forth the frontiers of science and contribute our fair share to the collective knowledge of Man; in this quest lies dignity for our countries and ourselves. I hope, my children, that one day the two of you will also walk this trail and do your part to mark out a path for those who follow.

I am counting the days until I see you. Until then, all my prayers are with you.

Your loving father

STRING THEORY: GRAVITY JOINS THE PARTY. IS IT ALL COMING TOGETHER AT LAST?

Cambridge, MA

March 2010

I can feel the space in the room expand as people shuffle out, books in hand. Almost all of them are discussing the lecture we just heard and most of them, as they head out the door, pause to shoot me one last over-the-shoulder glance. Even without looking up, I can make out the slightly confused expression on their faces and I know just what it indicates. The mind of a graduate student is a strange and wondrous thing; were I not constantly wrestling to control one myself, I might even find it amusing how all of us obsess about the same things, day in, day out, until eventually new thought processes begin to lay themselves over the existing vastly different patterns in our individual minds. We retain our uniqueness in other avenues of life, but in certain distinct ways, we are all the same, "marked" by this experience; even with my very limited knowledge of Star Trek, this reminds me of the assimilation by the Borg – suddenly, being only dimly aware of how it happened, we find ourselves part of the *Graduate Student Collective*, able to read each other's minds simply because those thoughts are our own as well.

Right now, for instance, behind those puzzled looks shot my way, I can sense a whole sea of insecurities. "Why is she still writing? What is there to write about? What did I miss? …" I want to reassure them somehow, but most of them I don't know well enough to respond to their unasked questions. I guess the only thing to do is to not make a habit of lingering here after class – I will simply take my writing elsewhere. But today, I do not want to break the flow of my thoughts even for as long as it would take me to stuff my laptop in my backpack and settle down on the grass behind Jefferson. Today, I just want to keep typing; letting my fingers rain on the keyboard while my thoughts manifest themselves on the screen … I have waited for this day long enough. I want to feel, and remember, every part of it.

I glance over my notes, pressing the Save button once again, just to be sure. Even as my eyes skim the heading – Physics 287a: *Introduction to string theory*, a smile creeps over my face. I have to stop doing that or people will think I'm insane! I caught myself smiling a couple of times during the lecture as well … of course I immediately toned it down to what I hope looked like the kind of smile sophisticated people engage in during seminars – a polite acknowledgement of interest, an academic nod of sorts. I'm afraid, however, that my smile probably looked instead like what it was - an expression of delight, an exuberant shout of "I made it! I'm here!"

I was still in my early teens when popular accounts of string theory flooded the popular consciousness. On my fourteenth birthday, my parents gave me "The Elegant Universe" by string theorist Brian Greene as a birthday gift. "Calling it a cover-up would be far too dramatic" it began, "But for more than half a century … physicists have been quietly aware of a dark cloud looming on a distant horizon." Straight away, I was hooked.

I went on to read, in that book and others, how the two fundamental pillars of theoretical physics, general relativity and quantum mechanics, are incompatible with each other. General relativity takes for itself the domain of the large, the heavy – it rules over galaxies and dictates star formation. Quantum mechanics on the other hand, reigns undisputed over subatomic scales. Thus far, the two have been able to co-exist simply because they do not often come head to head; very tiny things are seldom very heavy, so for most objects or phenomena in the universe, either one or the other theory will suffice.

While this is a working compromise, it is clearly not a solution. Even though it is not the norm, there *are* some extreme cases where things are small yet heavy; the early universe for instance, or black holes. In order to understand how nature behaves in these situations, aspects of quantum mechanics need to be merged with gravity. This is what string theory claims to do.

But there's more. String theory not only unifies quantum mechanics with general relativity, but in fact goes a step further. From its equations arise all four fundamental forces, as does matter. It appears that strings tie together all that we know. In the intellectual and aesthetic legacy we have inherited, it just doesn't get any better than this: one ultimate equation that contains the universe. This is the ideal the ages have aspired to. In the words of Socrates "It seemed to be a superlative thing to know the explanation of everything, why it comes to be, why it perishes, why it *is*". By reducing everything we know, matter and forces alike, to a single common root, string theory appears to contain the "final answer".

The essential idea is so seductively simple: the universe is created from strings. In retrospect, it gives me the shivers to think how obnoxious I must have sounded when I went around in high school saying "The whole point of string theory, is that there's no such thing as a point!" Awful. Pretentious. But kind of true ... It turns out that the quarks and leptons which we had considered to be point-like objects with no further structure, were not quite that; had they been Russian Matrushka dolls, they'd be the "last doll but one". One last tiny secret still lay hidden inside – the string; and oh the things this string could do! For starters, there's just one of it ... there are no families, like those of the leptons, it doesn't come in flavors, as the quarks do; the string carries no baggage. It just *is*; which, by the way, takes some getting used to. For the first time ever, we are faced with something that is so truly fundamental that it makes no sense to dig deeper and ask what the string is made of. The string is made of nothing, it is what it is – everything *else*, however, is made of strings. All the particles we are familiar with are just manifestations of the string vibrating in different ways. One little restless string contains within itself these infinite possibilities; it can be anything, it can be everything.

I have read this analogy so many times over the years that it is beginning to sound a little tired, but I must admit that the first few times I heard it, I could not help but thrill to the idea that, just as a single guitar string can be plucked in various ways to yield different sounds, every unique particle that makes up our world is in fact a distinct "note" played on the fundamental string; the basic string is the same, it is only the sound that changes.

To each particle, or "mode of vibration" of the string, corresponds a unique – quantized – energy determined by its oscillation. Luckily it isn't that hard to understand why this should be so. Picture the string as a little loop of energy, kind of like an infinitesimal rubber band, perpetually oscillating at the speed of light. Even intuitively, it is obvious that – just like the standing waves we read about in high school – only those oscillations that complete their cycles in going around the string will be stable; all other oscillations will interfere with themselves and decay. In other words, these holding patterns are possible only when the length of the string is an integer multiple of the wavelength of oscillation. Clearly, there are only a discrete number of possibilities (½, ¼ etc) and hence our string has a discrete set of notes in its repertoire. These "notes" give rise not just to the matter particles (i.e. the quarks and leptons) but also to the force-carrying particles (the gauge bosons), and so string theory subsumes not only all of matter but also all the known forces.

One of the reasons that a quantum theory of gravity had been impossibly hard to write down was that, in the formalism of quantum field theory, which had been so successful in describing all three of the other fundamental forces, space-time was taken to be a fixed background; it was merely the stage on which the play was acted out and, in true Shakespearean spirit, we assumed that "the play's the thing" and neglected gravity. But, in doing so, we apparently inflamed the gods of gravity. As Einstein had warned us, gravity is a living thing – it is the response of a dynamic spacetime to the dance of matter upon it. In any satisfactory theory of quantum gravity, spacetime cannot be relegated to the background; it takes its rightful place as a lead actor. And since string theory treats gravity with appropriate deference, we find that spacetime too is made of strings. *Matter, forces, spacetime* … all the wonderful chaos of this crazy confused world around us is a melody played on strings.

No work of fiction could have a wider scope than string theory ; no conflict between two characters could possibly be more dramatic than the battle between general relativity and quantum mechanics … The primal instincts within me responded to this, as much as the civilized intellectual part of me. Against this united onslaught, I didn't stand a chance.

In the decade between then and now, I devoured everything I could get my hands on. Luckily, many books have been written for the layperson, some by the "living legends" of string theory. I realized, straight off the bat that an equation which *contained the universe* would clearly be extremely dense. Every real masterpiece deserves to be read in the original language and I knew that a proper "reading" of string theory would have to wait until I had learnt the necessary mathematics. I quickly found out that there was a lot of it. Calculus, which gave my classmates nightmares at the time, was not even the beginning … years of college mathematics lay ahead. Along the way, algebra would become abstracted and eventually combined with geometry; I would learn to study objects of various un-visualizable shapes, in more dimensions than any sci-fi author would dare to mention … Clearly, this was going to take a while. So, even though I knew that any non-mathematical description of string theory would necessarily be an extremely stripped down version of the truth, I kept reading.

As I sat in class today, listening to Prof. Cumrun Vafa outline the course, I felt as if finally, I was coming in to my own. All those years of wrestling with atomic orbitals, statistical mechanics, circuits and semi-conductors have finally paid off. I now get to be initiated into the mysteries of string theory by one of its high priests.

I still dwell, both literally and metaphorically, on the "ground floor" – in the graduate student commons, to be precise. Only next year, after my coursework is over and research begins, will I get to move up to an office on the fourth floor where the theory group goes about their business, teasing out fundamental laws from squiggles on blackboards. This is the plane to which I aspire, but I am not one of its denizens yet; walking up to class today, I took it all in – the golden yellow walls, the exposed brick, the bronze bust of Einstein – they all become part of the mystique and the promise. Among the researchers who call this floor home are famous world-class physicists; the names on these doors are repeated in textbooks, both in the pages and on the covers. I have heard them mentioned reverently for years now, and even though I have seen most of them on and off, in colloquia and other departmental gatherings, I am not really familiar with them yet, so for me, some – okay most – of the awe of their reputations clings to them still … which is partly why I am so thrilled to be attending a course taught by Vafa.

He started with an overview, a brief sketch of what is to come this semester and I could almost hear Murray Gell-Mann 's words in my head "If a child grows up to be a scientist, he finds that he is paid to play all day the most exciting game devised by mankind". My one reservation about this course, or actually about seeing string theory "from the inside", was that the mystery of string theory would vanish and leave behind only the mechanics, but that turned out to be a completely unfounded worry. When Prof. Vafa mentioned the "rich and rigid structures" we would encounter this semester, it was impossible to miss the passion in his voice; even though it was all so basic and he was going through this for the umpteenth time, his joy was still palpable – and in fact, infectious.

He spoke about things with such ease that it was clear they were real to him. Whether he sketched a seven dimensional sphere on the blackboard, or spoke about the strong coupling limit of Type IIA superstrings, his descriptions were so effortless that it was obvious he lived among these things; he moved around them with such grace and familiarity, he saw them almost as if they were tangible and in those moments, he made me feel as if I saw them too.

At the end of class, Professor Vafa told us that next time we would get down to the nuts and bolts of *supersymmetry*. "This is an involved and sometimes unexciting discussion," he warned, "but supersymmetry runs the show, so it is not enough that someone somewhere knows how it works – you need to know it for yourself. At least once in your life, you need to work through the calculations yourself". Now, I was never one to take things apart; I had no need to see the innards of either machines or electronic gadgets. Supersymmetry, however, is something I've been curious about for a while; I can't wait to take it apart – and learn how to put it back together.

When I first came across this concept, I had already read about several unifications in physics; I had been excited by them all. It was wonderful the way Maxwell united electricity and magnetism and I was positively thrilled to discover that the weak force was literally like a long lost sibling of electromagnetism ; I loved these ideas, but – cool as I was – I took them in my stride. Supersymmetry, on the other hand, completely knocked me over. Digging out the family tree and finding two forces were related is admittedly incredible, but knocking down the boundary between matter and force … well, the audacity is almost unbelievable.

Until just a few decades ago, it seemed perfectly clear that all particles could be split into two camps: they were either *fermions* (which made up matter) or *bosons* (which carried forces). The two were so obviously different; they obeyed different rules (of statistics), they had different purposes … there was no possible way of confusing the two; and then supersymmetry came along, saying that fermions could "transform" into bosons and vice versa. This was mind-blowing stuff. If two objects are related by a transformation, the distinctions between the two melt away – they both appear to be different aspects of the same thing. It's like finding out Clark Kent and Superman are the same person; suddenly, you know more about him than you ever knew about either of his two personas – there are dots you can now connect that you couldn't before: that's what supersymmetry did to fermions and bosons.
As I read on, I realized that there was more to it. The universe around us today is not supersymmetric; bosons and fermions cannot switch identities merely by putting on – or taking off – a cape and glasses. Supersymmetry is conjectured to have been a feature of the early universe, but at some point along its evolution the universe made some asymmetrical choices and this symmetry *broke*.

String theory incorporates supersymmetry : the proper name in fact, is *superstring theory* … but no one says that anymore. In fact even the epithet "String Theory" is kind of outdated – the theory has evolved to include so much more, but somehow, perhaps for historic reasons, the name stuck.

Figure 2: Strings in the air?

Luckily, I belong to the generation that came of age well after the *second string revolution*. I did not have to wait with baited breath, wondering if string theory was perhaps *too much of a good thing*. In the early 90s, for a while, the fate of string theory seemed somewhat doomed. There turned out to be five copies of the theory that claimed to unite everything. It appeared that there were several, perfectly consistent, ways in which string theory could be written down. These "versions" were given the unimaginative names of Type I, Type IIA, Type IIB, $SO(32)$ Heterotic and $E_8 \times E_8$ Heterotic string theories. Talk about an embarrassment of riches! The question now became: *which* theory did Nature choose, and why.

Thankfully, for all concerned, the theory was rescued in a single talk by Ed Witten at the annual Strings conference in 1995. Witten had done much important work before then, and he has done much since, but that day, in what was probably his most dramatic moment, he showed that the five theories we thought of as distinct were in fact merely different aspects of the same thing. The five string theories, it turned out, are connected via a web of *dualities*. This led to many repetitions of the story of the blind men and the elephant (the one in which the five blind man each touch a different part of the elephant and insist the other is wrong until they realize that they are all describing a piece of the truth); the elephant, the over-arching theory, came to be known as *M-Theory* and all was well. Despite the fact that no one really knew what the 'M' in M-Theory stood for[14], it appeared once again, that there was just *one* contender for the title of "Theory of Everything", which is just how it should be.

In the decade and more that has passed between then and now, much work has been done. Sometimes in baby steps, sometimes in leaps and bounds. One major breakthrough was the discovery of "*p*–branes ", the hyper-dimensional "membranes" that lay hidden in the equations until we started asking the right questions. This made it clearer than ever before that "string theory " was a misnomer, but until someone has a better idea, perhaps the name will do ...

One of the problems of studying string theory is that it has attained celebrity status. Where there has been positive press, there have also been rumors. As a result, I am often told by "well-wishers" that I am making a risky bet; that now is not the time to start out. "What if they find nothing at the LHC?" I am asked. "What if by the time you graduate, string theory is dead? What will your degree mean then?" My friends ask me what good "all these hours spent pouring over arcane mathematics" will be, "if string theory turns out to be wrong". Most of the time, I just respond politely saying that I really don't think that will happen, and in any case, I do what I do because I love it – not many people can make the same claim. On the inside though, the repeated question sometimes enflames me a little. How can string theory be *wrong*? At times I feel like responding "A truth can't be wrong", but I know that is only going to draw me into an endless – and perhaps heated – debate; the question actually isn't whether the theory is mathematically consistent but whether it is physically realized: *does it describe Nature*?

Admittedly, there is no single test one can perform right now, to prove – or disprove – string theory. This is a source of great frustration for most, and understandably so. Physics has a reputation of being a science that can be tested and verified through experiment and currently that does not seem possible with string theory ... but the fact is that the theory isn't "done" yet. It is still a work in progress.

This morning I was reading about a beautiful "art experiment" that is being carried out in honor of the LHC, at the Niels Bohr Institute in Copenhagen. There is no doubt that the LHC is a phenomenal, historic achievement. It is a triumph of engineering, international collaboration and a commitment to the search for answers. By recreating incredibly high energies, of the range of few TeV[15], the LHC gives us access – for the first time, and in a controlled manner – to conditions like those that existed in the early universe. Perhaps to highlight the beauty and importance of what is happening there; the façade of Bohr's Institute wears for now a "Colliderscope". This is the name given to a light installation, consisting of nearly a hundred diodes which with an ever-changing stream of light, faithfully reconstruct the events taking place inside the accelerator at CERN. This information is transmitted live, from Geneva to Copenhagen and the tracks made by particles at the LHC are mimicked by the changing patterns according to which the lamps light up and fade away, the intensity of their glow denoting the energy and mass of the particle they represent. And so, in this uniquely artistic tribute to experiment, instead of the numbers and curves usually associated with particle collisions, the Colliderscope presents an immediate and accurate visualization of the LHC data in a giant, dynamic light sculpture[16].

---

[14] The official tag-line reads: "M is for Magic, Mystery or Matrix, according to taste", but it could equally well stand for Membrane.

[15] Teraelectronvolts. The *electronvolt* is a unit of energy traditionally used in particle physics. Tera=10^{12}. Also used is the Gigaelectronvolt: Giga=10^{9}.

[16] As of the time of this writing, there are some videos on *YouTube* showing the Colliderscope.

The physics community has high hopes from this "particle smasher". We hope that we will finally get to see the Higgs boson – the fabled particle predicted by the standard model that goes around bestowing mass upon others; perhaps – fingers crossed – we might catch a glimpse of supersymmetry in action and gain some insight into the mystery of dark matter ; in principle, it is not impossible even to observe some of the extra dimensions string theory got such bad press for! Even as I write this, several string theorists are working on phenomenological models that link the elegant equations we are so in love with, to the crazy world we live in. Some of these models might even make predictions that can be tested at the LHC.

But until concrete measurable predictions are made, and tested, I will continue to believe in string theory and not just because of my innate obstinacy. One of the reasons for my faith is that across the centuries, throughout the history of physics, paradoxes have always led to progress. Whenever two seemingly different paradigms are brought together, the whole has always been greater than the sum of its parts; and if the paradigms appear to contradict each other, their unification has yielded tremendous benefits.

In the case of string theory, these contradictory paradigms are two of the most fundamental and most rigorously tested theories known to man; moreover, between them, they span the entire gamut, from small to large, light to heavy, nothing seems to have fallen through the cracks. It is only natural then to suppose that if we can bring together these two theories, which currently stand at opposite ends of a long line, we will finally have closed the loop. Of course this is just a "belief", but even though it is not often stated in so many words, such beliefs are woven into the very fabric of science. It is just a belief, for instance, that the universe is *comprehensible* in the first place. Everything we do is based on the *belief* that Nature can be mathematically described.

String theory, many features of which developed as a mathematical curiosity, is giving us things that fit our world; things we had not even asked for – so how wrong can it be? Moreover, while there is no absolute proof – and that may be the case for a while yet – there are some indications that we are on the right track. There is, for example, the phenomenal success with which string theory answered certain puzzles related to black holes. About three decades ago, Stephen Hawking came up with the now famous *information paradox*. Using quantum mechanics, he showed that particles could "boil off" the surface of a black hole, eventually causing the black hole to evaporate and all the information it contained would be irreversibly lost.

This was anathema for physicists, but no one could really find a way out ... until about twenty years later, when Andrew Strominger and Vafa used string theory to measure the information contained in a black hole. Roughly speaking, their calculations showed that black holes contain strings which store information in such a way that it is never completely lost. Their string theoretic explanation reproduced the standard, well-known result for the black hole's entropy and yet had none of the contradictions or discrepancies that arose in earlier attempts when either general relativity or quantum mechanics was used to describe black holes. This success was taken by many to be a clear sign that string theory was on the right track.

Another fact that gives me hope is that until relatively recently, people who searched for unification did so in solitude. Even an icon like Einstein, when he decided to pursue this quest, had to do so alone. But over the years between then and now, more and more people joined the fray. Today, string theory groups have popped up in the best universities the world over and some of our planet's brightest people are working on this problem, *together*.

The past decade has made me a bit of a popular science junkie. Even now when I am immersed in technical details, I still read books for the "educated layperson". I find this fun, not only because I get an insight into how different people process the same thing but also because it helps me keep my perspective; every once in a while when I get a little lost in calculations, these books force me to take a step back and think of the bigger picture.

Regarding this whole business of whether a theory is right or wrong, one of my favorite science writers, K. C. Cole, has this to say "Right ideas seem to be those that lead to further investigation, to whole new categories of questions, to an even more passionate quest for knowledge. Right tends to open our eyes, wrong tends to close them".

I could not agree more. And therein lies another one of my reasons for believing in string theory. Wrong wants to draw sharp boundaries and encircle knowledge; make it dark and final. Right is aware of what it misses. It is exultant, yet humble in its acknowledgement of omissions. Despite the tall claims that sometimes reach the outside, the people who actually stand at the front-lines are remarkably aware of all the missing pieces and the tantalizing mysteries that remain to be absorbed into this magnificent structure; they routinely get together and list all the things that are not yet explained by the existing framework and try to figure out ways of extending it. In this, perhaps, might lie the salvation of String Theory. Already this theory has considerably evolved from what it used to be when it started out. From a purely perturbative theory of strings, the net has expanded to include objects that are both non-perturbative and higher dimensional. The final theory might end up looking quite different from what it started out to be, just as an adult sometimes is unrecognizable from baby pictures, but I find it hard to believe that the beginnings are not there. Edging towards M-Theory, whose very name quantifies our ignorance, we celebrate what we know, but are quite conscious of what we don't. And if at any time we find ourselves getting arrogant in our knowledge, all we need to do is remind ourselves of Hawking 's question: "Even if there is only one possible unified theory", he says "it is just a set of rules and equations. What is it that breathes fire into the equations and makes a universe for them to describe?"

CHAPTER 2

String Theory and the Failure of Unification

Peter Woit

Department of Mathematics, Columbia University

Abstract: The subject known as *string theory* by now has a forty-year long history, during which it has evolved into a collection of various active sub-fields in physics and mathematics. With a literature of several tens of thousands of papers, many using highly abstract and specialized mathematical techniques, the degree of complexity involved is hard to over-emphasize. This situation makes it very difficult to evaluate the state of the field and reach conclusions about which ideas have worked out and which haven't. This article will focus on one particular aspect of string theory, the hope that it might provide a unified theory including both general relativity and the Standard Model. The argument will be made that while this is what initially led to the explosion of interest in string theory back in 1984, by now there are solid reasons to believe that attempts to get unification along these lines are doomed to failure. For a more extensive version of the discussion here (with much more in the way of references), see the author's 2006 semi-popular book [1]. In recent years, one segment of the string theory community has reacted to the problems with string theory and unification by conjecturing the existence of a *string theory landscape*, the existence of which would make it impossible to get testable predictions out of string theory. This move away from the standard scientific method, motivated by an unwillingness to give up long-cherished hopes, holds significant dangers for theoretical physics should it become conventional wisdom.

BEYOND THE STANDARD MODEL

By 1973, all the elements of what is now known as the Standard Model were in place. This was often described in the 1970s as a unified theory, since it brought together the electromagnetic, weak and strong interactions with a uniform description of forces and matter particles. The one known force left out was the gravitational force, which has to this day resisted attempts to find a convincing way of understanding it in terms of a quantum gauge field theory.

One obvious target for researchers trying to find a more fully unified theory than the Standard Model is the question of how to come up with a quantum version of Einstein 's general relativity. A major problem with any such research program is that simple estimates of the size of quantum gravitational effects make them unobservably small for any physical system we can access experimentally. A massive experimental effort is now underway that has some hope of being able to detect classical gravitational waves, but the gravitational quantum particles; *gravitons*, are likely to remain undetectable for a very long time. Given this situation, any proposal for a quantum theory that only deals with gravity appears to have no hope of being experimentally testable. One might hope that the criterion of uniqueness would save the day, with only one possible consistent method for quantizing general relativity. Unfortunately, it seems all too likely that this is not the case, with advocates of *loop quantum gravity*, *N*=8 *supergravity*, *asymptotic safety* and *string theory* all making a plausible case that their research program shows the way to one or more such consistent theories[1].

In the absence of directly relevant experiments, one way to make a convincing case for a quantum gravity theory would be to find a unified theory based on principles that had gravity as one consequence, but also had other, more readily testable, distinctive consequences. The hope that string theory might be able to do this was one of the main things that got people excited about it around 1984.

The Standard Model is based on a branch of mathematics known as *gauge groups*. These are collections of symmetries, e.g. mathematical transformations, under which a certain theory of physics is invariant. These groups are usually given rather non-descriptive names. For example, the symmetry group of Maxwell 's theory of

[1] *Editor's note*: Loop quantum gravity and asymptotic safety are discussed in detail in "*Science Fiction of Everything*" and "*A Vision of Quantum Gravity*" respectively.

Moataz H. Emam (Ed)

electromagnetism is called $U(1)$. Collectively, the group structure of the Standard Model is known as $SU(3)\times SU(2)\times U(1)$. In the Standard Model, the $SU(2)$ symmetry is spontaneously broken, i.e. it is not observed directly in our everyday low-energy environment, but is seen only in high energy experiments.

The list of questions that any more unified extension of the Standard Model might be expected to answer is a relatively small and well-known one, and goes roughly as follows:

- Why $SU(3)\times SU(2)\times U(1)$ and why do the corresponding coupling constants (i.e. the strengths by which the fields interact) take the values that they do?
- Why do quarks and leptons behave under $SU(3)\times SU(2)\times U(1)$ the way they do?
- What causes the spontaneous breaking of the $SU(2)$ gauge symmetry?
- Why do quarks and leptons have the observed masses and other properties?

The only known successful answer to the first two questions on this list comes from grand unification; the idea that there is some larger symmetry group that breaks down to $SU(3)\times SU(2)\times U(1)$, i.e. has those (smaller) groups as constituents. The simplest version of this idea uses the group $SO(10)$ and organizes the known quarks and leptons so as to transform accordingly. This still leaves the question "why $SO(10)$", and introduces a new spontaneous symmetry breaking problem: what breaks the $SO(10)$ symmetry to $SU(3)\times SU(2)\times U(1)$?

The best answer we have to the last two questions is a rather unsatisfactory one. It involves the introduction of a new elementary scalar field, the *Higgs field*. This introduces problematic high-energy behavior into the theory. There are no symmetries that constrain the Higgs coupling to matter fields, so one ends up with a moderately large number of extra undetermined parameters, the existence of which leaves the values of matter masses and other properties as much of a mystery as ever.

THE FIRST SUPERSTRING REVOLUTION

Grand unified theories (GUT) were first studied in 1974, very quickly after the Standard Model came together in 1973. The simplest of these came into conflict with experiment as they predicted the unobserved decay of the proton. Supersymmetric extensions of GUT models evaded this problem and more accurately allowed for coupling constant unification, but added a whole new set of difficulties associated with the necessity of breaking the supersymmetry [2]. By 1984, the lack of any convincing progress on getting beyond the Standard Model had begun to become discouraging.

For several years before then, a small number of physicists had been investigating string theories and their supersymmetric versions, motivated by the idea that they might provide a way to achieve unification of the Standard Model with gravity. Quantization of string theory was a difficult subject, with the quantization procedure producing *anomalies* that violated the symmetries of the classical theory. It became clear after a while that some of these problems vanish in a specific critical spacetime dimension, which was 26 for the non-supersymmetric Bosonic string, 10 for the superstring. Attention focused on the 10-dimensional superstring, with the hope of somehow extracting from it a supersymmetric GUT, which would be an extension of the Standard Model. A remaining technical problem was the possibility of an anomaly appearing in the gauge symmetry of such a theory. During the summer of 1984, Michael Green and John Schwarz showed [2] that this problem disappeared for a specific choice of gauge group, the group $SO(32)$, which attracted the attention of Edward Witten at Princeton, who began to work intensively on the problem of how to use the superstring to get a viable unified theory.

Within a few months Witten and others at Princeton and elsewhere had come up with various proposals for how to do this. The Green -Schwarz anomaly cancellation mechanism in a sense provided an answer to the question of *Why a specific gauge group*, but the answer required that the group should be $E_8\times E_8$. To get a viable GUT group like the much smaller $SO(10)$ requires the addition of yet another set of extra structures, the ones necessary to break the larger symmetry down to $SO(10)$.

---

[2] Since supersymmetry is not observed in our low-energy experiments.

Similarly, it is also an anomaly cancellation calculation that gives the critical spacetime dimension of the superstring (ten). This was not the answer one wanted (four), so the search began for ways of making the other six dimensions unobservable. The most popular way to do this involved taking the extra six dimensions to have a geometric shape known as a *Ricci-flat Kähler manifold*, also known as a *Calabi-Yau manifold.* With such a choice one could arrange for the low-energy field theory limit to have four-dimensional supersymmetry and thus give the sort of supersymmetric GUT model considered to be the most promising extension of the Standard Model.

This speculative unified framework quickly drew a lot of attention, with many particle theorists deciding to start studying string theory and working in this area. However, from the beginning, there was no especially compelling evidence for such a unified theory. The proposed string theory framework was powerful enough to answer some questions the Standard Model could not address (Why a specific number of spacetime dimensions? Why a specific choice of gauge group?), but it gave not the observed answers, rather much larger ones. As a result, one could presumably embed the Standard Model in a string model but, suspiciously, all the questions one wanted answers to just got transformed into another question: why this specific embedding?

Among the various choices to be made, the first was the choice of a Calabi-Yau manifold. These come in families, of a dimension computable by the methods of algebraic geometry. At first very few examples of such manifolds were known, but it quickly became clear that the number of families was at least of order 10^5. The question of whether the number of Calabi-Yau families is finite remains to this day an open question in algebraic geometry.

Each Calabi-Yau family has a dimension[3], often of order of magnitude 10-100, and the parameters specifying elements of the family are known as *moduli parameters*. The low-energy physics will depend on how one chooses these parameters. One would like to take the moduli parameters to be dynamical variables, but then one needs to understand what dynamics governs them. One needs some sort of *moduli stabilization* mechanism to pick out specific values. These effectively must be non-trivial, otherwise the moduli parameters will give new massless fields and thus new long-range forces in the low energy theory[4]. The mechanism must also not allow the size of the Calabi-Yau to be zero or infinity. Despite intensive effort during the late 1980s and early 1990s, no solution was found to the moduli-stabilization problem.

BRANES AND M-THEORY

String theorists often describe the status of the subject with phrases like *We don't know what string theory really is.* What they mean is that no well-defined fundamental definition of the theory exists. Instead we have various theories that are supposed to be approximations to the fundamental theory in certain limits, together with various consistency conditions that are assumed to hold in whatever the fundamental theory might be. Quantization of the classical string theory gives something roughly analogous to quantization of the theory of a relativistic point particle. From such a point particle theory, one can build up a more complete theory based on the so-called *perturbation technique*[5], but this is only an approximation to something much more fundamental, a quantum field theory. For some aspects of the Standard Model such an approximation is useful and perturbation theory provides a powerful calculational technique, for others (e.g. the low-energy limit of QCD; the theory of the strong nuclear force) the perturbative approximation is useless.

The most straightforward way to come up with a truly fundamental nonperturbative version of string theory is to try and emulate the point particle case, and define a *string field theory*, but so far this has not led to a useful, consistent theory. The year 1995 saw the advent of a series of related ideas that provided a new direction for the investigation of what nonperturbative string theory might be. One aspect of these ideas involved various duality transformations relating the different known ten-dimensional superstring theories. The name *M-theory* was given to the conjectural underlying nonperturbative theory that was supposed to reduce to the superstring theories in various limiting cases. There have been various more or less concrete proposals for what M-theory actually is, but none of them are fully satisfactory.

---

[3] *Editor's note*: The dimensions of Calabi-Yau manifolds discussed here are *not* their spacetime dimensions, but rather more intrinsic ones.
[4] Forces that are unobserved in nature.
[5] *Editor's note*: The perturbation technique is discussed in more detail in "*Science Fiction of Everything*".

One aspect of the conjectural M-theory story involved the expansion of the class of objects considered to include not just strings, but branes. As strings move they sweep out a two-dimensional space, the so-called world-sheet of the string. In the perturbative string formalism, a string theory is essentially given by a choice of an appropriate two-dimensional quantum field theory that lives on the world-sheet. Such field theories have a rigid and beautiful mathematical structure, and during the late eighties much effort went into their classification. What Polchinski and others discovered around 1995 was that one should study those field theories together with a specific choice of boundary conditions. Such conditions were called *branes*. A basic example of a brane would be boundary conditions that fix the ends of an open string to lie on specified spacetime points, and the term *brane* sometimes refers to the collection of such points – see (Fig. **1**).

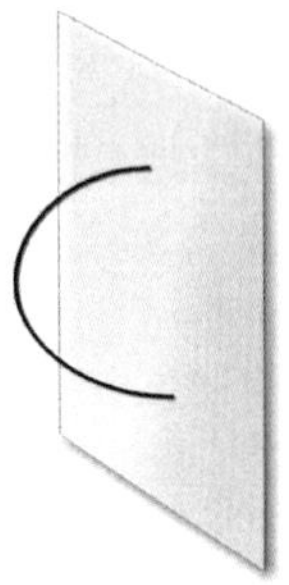

Figure 1: A simplified view of a brane, in this case a 2-dimensional surface, as the boundary condition on the ends of a string. The string can move any way it wants so long as its endpoints stay on the brane.

While M-theory and branes opened up a host of interesting questions related to string theory, they in some sense made the string theory unification situation much worse. M-theory was conjectured to be something that unified the known string theories in a unique (but still unknown) theory, but it required the introduction of a whole new set of structures (the branes), which could be chosen in many different ways. These new choices led to a wide variety of new possible ways to embed the Standard Model in string theory. Instead of getting rid of six dimensions using a choice of invisibly small Calabi-Yau manifolds, one could instead just arrange for our four-dimensional spacetime to be a brane inside ten-dimensional space. By appropriate choices of branes, with appropriate relative positions and intersections, one could produce an even wider variety of ways to embed the Standard Model in string theory than before. At least initially, these extra choices did not seem to help with the moduli -stabilization problem.

THE LANDSCAPE

In 2003, finally a proposal [3] appeared for how to solve the moduli stabilization problem, using a combination of branes and choices of certain properties of the Calabi-Yau manifold. In this proposal, things could be arranged so as to make a meta-stable ground state of string theory, one with a lifetime that can be arbitrarily long. In this state, the moduli parameters would take on fixed values, and in principle one could extract an effective low energy quantum field theory from the set-up, and see if it reproduced the Standard Model. The graph of values taken on by the effective theory as one varied moduli parameters (as well as other choices such as branes and fluxes) came to be known as the *string theory landscape* [4].

Unfortunately, this was a rather Pyrrhic victory, as it became clear that typically the effective theory would have not one form, but an exponentially large number of them. An estimate of 10^{500} such forms is often quoted (for a particular choice of family of Calabi-Yau 's). There is an inherent technical problem with the idea of performing an exponentially large number of calculations (any one of which is even now much too difficult to carry out). In addition, the lack of any highly constrained symmetry structure to these calculations means that one expects to get a fairly random set of numbers as results (and calculations in simplified examples bear out this expectation). As a result, one expects that in this framework string theory unification is an idea that lacks any of the sort of predictivity that would allow it to be compared to experiment and evaluated in a conventional scientific manner.

Some string theorists have gone so far as to try and turn this failure into a virtue, arguing that the existence of string theory states with just about any possible value of the constants of nature provides an anthropic explanation of the particular laws of physics in our universe. The idea is that our universe is meta-stable and just one of an ensemble of meta-stable universes produced by transitions in a *multiverse*. The constants in our universe are not numbers that can be calculated, but the result of whatever history of transitions led from other universes to ours. The only prediction then possible about these constants is a statistical one, based upon combining the probability distribution of the emergence of life with the probability distribution generated by the transition process. Even if one could come up with a reasonable calculation of the first distribution, the complexity of the landscape makes a calculation of the second distribution impossible. Instead of such a calculation, proponents assume a flat distribution over the relevant energy scales. The calculation of the constants' probability distribution based upon string theory is then exactly the same calculation as one based on the model: "we have no idea what determines the constants, so let's take all values as equally likely". The problem with statistical predictions from the string theory Landscape is that what one gets is not distinctive of string theory and can't provide significant support for it.

It may very well be that we live in a multiverse, with the fundamental constants historical accidents that can only be statistically determined. But the conventional standard of scientific evidence requires that to be considered scientific, a theory of such a multiverse must make convincingly distinctive and falsifiable predictions. String theory fails to live up to this standard, and it is disturbing to see attempts made to give up on conventional scientific standards in preference to giving up on a failed theory.

STRING THEORY'S PRESENT AND FUTURE

For much of the past decade, many string theorists have abandoned work on string theory unification, preferring to investigate some of the remaining poorly understood aspects of the underlying theory itself. By far the most popular such effort is the study of AdS/CFT and other conjectured dualities between gauge theories and string theories (or at least their low-energy gravity limits). Such dualities provide a conjectural answer in some regimes to the question of how to define string theory nonperturbatively, and thus a possible approach to certain nonperturbative questions in quantum gravity theories. They also provide a return to string theory's roots, to its use as an approximate calculational method for studying certain strongly coupled quantum field theories. A current hot topic is to try and do this for certain quantum field theories possibly relevant to condensed matter physics. This kind of study is now a very long way from string theory unification.

Hopes to revive string theory unification rest mainly on the idea that new data from the LHC or astrophysical experiments will change the current situation and provide some sort of positive evidence for a particular string theory construction. This construction might be sufficiently constrained so as to have some sort of predictive power, making testable predictions that could be checked. In the case of cosmology, it is hard to see how this could happen since the small number of new observables measured by next-generation experiments are not such as to provide any way to pick out one particular sort of string theory. For the case of the LHC, the hope is for a whole new sector of particle physics and picking out one particular supersymmetric low energy theory with one particular supersymmetry breaking mechanism. Given that specification of the supersymmetry breaking mechanism typically still allows a wide range of compatible string vacuum constructions, it seems unlikely that even this scenario would fix the predictivity problems of string theory. In any case, with the LHC starting up this year and finally providing an answer to the question of whether supersymmetry exists within a few years, we'll soon know if this possibility gets eliminated or not. If it is eliminated, it is very hard to see what the possible scientific basis might be for further research on string theory unification.

REFERENCES

[1] P. Woit, *Not Even Wrong*. Cape, 2006.
[2] M. Green, and J. Schwarz, *Phys. Lett.* B149, p117, 1984.
[3] S. Kachru, R. Kallosh, A. Linde, and S. Trivedi, *Phys. Rev.* D68, p046005, 2003.
[4] L. Susskind, *The Cosmic Landscape*. Little & Brown, 2005.

CHAPTER 3

Can String Theory Survive Complete Falsification?

Moataz H. Emam

Department of Physics, State University of New York College at Cortland

Abstract: The current pursuit of a complete theory of physics, one that incorporates both general relativity and the Standard Model , is wholly based on the power of pure thought. The lack of experimental verification and guidance has forced physicists to explore, and indeed expand, the vast mathematical landscape in search of ideas, one of which happens to be the staggeringly large edifice known as superstring theory. Many of us have specialized into some remote corner of this landscape with little or no hope of that corner ever making contact with reality. So certain questions naturally arise: Why are we doing this? Why is this theory of strings so attractive that, against all odds, it has survived for decades as an active field of physics research? In this article I ask this question, and discuss those achievements of string theory that *may* just be enough to allow it to survive, even if completely falsified as an explanation of nature.

INTRODUCTION

There is no doubt that string theory occupies a special niche in the history of physics, having achieved uniqueness in many unprecedented ways: It is a theory of quantum gravity that also aspires to be a theory of everything. It is the only physical theory that dictates the number of spacetime dimensions it needs. It is a theory that, beyond a few initial axioms, requires very little external adjustment, being almost fully self-contained from the get go. It is the only theory of physics with *absolutely* no experimental backing that has managed to not only survive, but also become "the only game in town" (to quote Nobel Laureate Sheldon Glashow). Furthermore, the theory itself, its evolution and the public's interest in it are completely unprecedented in the annals of theoretical physics. This last, in my view, is actually quite understandable. String theory, among other things, appeals deeply to what I often call the "Star Trek syndrome". The average sci-fi fan (and many scientists started out as such and still are, including this author) was perhaps surprised and quite excited to learn that higher dimensions, parallel universes and other exotic ideas have become part of orthodox science. Television programs such as Nova's "The Elegant Universe" hosted by Columbia University's Brian Greene and based on his book by the same name [1] have furthered the theory's popularity (The program is available online for free on PBS' website: *http://www.pbs.org/wgbh/nova/elegant/*). Not since the advent of relativity has any other theory of physics managed to capture the public's imagination to that extent. A few years ago when I would mention the words "string theory" to a non-physicist friend, the next question was inevitably: "What's *that*?" Today, this is no longer the case. Many non-physicists have not only heard the term, but are usually familiar with what it is, if only in a broad sense, and the numbers are continually on the rise. String theory has become so well known that many people actually take for granted that it is a correct theory of nature.

Of course, no one knows that for sure yet. This misconception, however, has extended even to physicists. Although when pressed, most, if not all, will tell you that string theory is still technically in the hypothesis stage, many do act and talk as if it were confirmed that it is a correct theory of the universe. This attitude has triggered a lot of criticism. One of the most vocal of string critics is Lee Smolin of the Perimeter Institute, though he himself has passed through a string theory phase. In his book [2] Smolin points out that this "faith" in the string hypothesis has affected the physics community in a negative way. In many cases, he argues, post-doctorate researchers and junior faculty are hired *solely* on whether or not they are working on string theory, and particularly on those parts of the theory that are fashionable at the time, resulting in the disproportionate inflation of the number of string theorists, as compared to physicists working on other approaches to quantum gravity [3][1]. While this author may not be particularly suited to judge Smolin's arguments on their merits, it is clear that string theory does receive a lot more hype than it perhaps deserves if evaluated solely on its applicability to nature and connection to experiment, as every theory of physics *should* be evaluated.

[1] Some of which are discussed later in this book.

IS STRING THEORY EVEN A SCIENCE?

Perhaps it is hard for many of us to admit that it is indeed a fact that string theory has so far failed to conform to the very *definition* of a scientific theory. In his classic work [4] famed science philosopher Karl Popper lists certain criteria that a scientific theory must satisfy. These are (in his own words):

1. It is easy to obtain confirmations, or verifications, for nearly every theory - if we look for confirmations.
2. Confirmations should count only if they are the result of risky predictions; that is to say, if, unenlightened by the theory in question, we should have expected an event which was incompatible with the theory – an event which would have refuted the theory.
3. Every "good" scientific theory is a prohibition: it forbids certain things to happen. The more a theory forbids, the better it is.
4. *A theory which is not refutable by any conceivable event is non-scientific*[2]. Irrefutability is not a virtue of a theory (as people often think) but a vice.
5. Every genuine test of a theory is an attempt to falsify it, or to refute it. Testability is falsifiability; but there are degrees of testability: some theories are more testable, more exposed to refutation, than others; they take, as it were, greater risks.
6. Confirming evidence should not count except when it is the result of a genuine test of the theory; and this means that it can be presented as a serious but unsuccessful attempt to falsify the theory.
7. Some genuinely testable theories, when found to be false, are still upheld by their admirers - for example by introducing *ad hoc* some auxiliary assumption, or by reinterpreting the theory *ad hoc* in such a way that it escapes refutation. Such a procedure is always possible, but it rescues the theory from refutation only at the price of destroying, or at least lowering, its scientific status.

"In other words", he goes on to say, "The criterion of the scientific status of a theory is its falsifiability, or refutability, or testability". It is interesting to note that string theory fails, quite miserably, to meet Popper 's conditions. Smolin makes this very argument in his book mentioned above, as well as another notable string critic; Peter Woit, in his book [5], articles [6, 7] and on his own website [8][3]. One might argue however, that this is a temporary situation. Eventually, one hopes, either technology will catch up with string theory and allows us to examine the mandates of the theory directly in the lab or someone will find a way to rigorously test the theory using current technology. This hope is what keeps string theory on the list of scientific theories, saving it from the fate of astrology and creationism. It is however a serious drawback that the string critics will, justly, continue to point out.

SO WHY DO STRING THEORISTS EXIST?

One often hears that, as unscientific as it may sound, the structure of string theory is aesthetically attractive, and that it would be a "shame" if Nature had not picked such an elegant structure to use as the basis of our universe. The general "criterion" of judging a theory of physics based on its mathematical elegance is sometimes attributed to famed theoretical physicist Paul Dirac , but there is no general consensus on how it can be defined; i.e. how can one, in fact, decide if a theory is "mathematically beautiful". Furthermore, unless you are an expert who has spent a considerable amount of time studying the theory's details, this elegance will not be immediately obvious. It is quite hard, in fact, to transfer this feeling to the non-expert[4]. One factor that has contributed to this theoretical beauty is the fact that string theory, compared to say the Standard Model , rarely requires external adjustment. Based on some initial axioms (the string nature of point particles, the existence of supersymmetry etc), the entire structure follows naturally with very little tweaking. Even fundamental properties, such as the number of spacetime dimensions required, are set by the theory's mathematics, something that no other theory has ever done. Usually theories in physics are set, from the start, in the four spacetime dimensions of our world. String theory *demands* its preferred

---

[2] Emphasis mine.

[3] As well as in this book in his article "*String Theory and the Failure of Unification*".

[4] Hard but not impossible. In fact, the aforementioned Nova program made a reasonably successful attempt at explaining the elegance of the string idea to the lay-person, let alone the non-string physicist.

number of dimensions, and, in fact, fails to be mathematically consistent otherwise[5]. Although there are, as noted, other approaches to quantum gravity , none of them have any claim on string theory's last and greatest appeal: it aspires to be a theory of *everything*; defining all matter and forces, even possibly answering deep questions about the nature of spacetime and the origin of the universe. I say "aspires" simply because there is as yet no rigorous proof that the physics we observe in nature is naturally contained in string theory, although the (mathematical) evidence leads most experts to assume it true. While there are some speculations on how to generate a theory of everything out of other approaches to quantum gravity , none of these speculations, in my meager understanding, even approaches the level of sophistication of string theory.

LANDSCAPING STRINGS

The general *hope* is that eventually we will have a complete nonperturbative (i.e. exact) quantum theory that reduces to the Standard Model plus general relativity in our low energy, perturbatively approximate world. Not only that, one would like this to happen in a *unique* way as well. It simply does not make sense to discover that *The* theory of everything is just one among several *possibilities*! Presumably there is only one universe, so there should be one theory explaining said universe.

Unfortunately, this is not straightforward. It has long been known that there are five supersymmetric string theories living in ten spacetime dimensions. For a while this seemed like an embarrassment of riches. You wanted *one* theory of everything, how about *five*? To the relief of most of us, it was later postulated that these five are not *really* different! They are now thought of as five perturbative (approximate) limits to a *single* unique theory in *eleven* dimensions; the hypothesized M-theory. While there is no rigorous proof of this, the evidence is of such strength that string theorists are content with this assumption, even though the full form of this M-theory is still unknown.

So, great! No more five string theories, just one theory of everything, M-theory, right? Not really! Those theories are still strictly higher dimensional. So there are too many spatial dimensions than we observe in nature. Once again theorists found a way out. This was achieved by reviving an early twentieth century idea known as the *Kaluza-Klein compactification mechanism*, which simply says that if you have too many unobserved extra dimensions in a given theory, all you need to do to "get rid" of them and fit your theory to the real world is to assume that they are wrapped on extremely small cycles, so small that they do not directly feature in any observable way in our much larger scales of existence. How can one visualize this? We can't, but it can perhaps be understood by analogy: Consider a two dimensional universe – as in (Fig. **1**), like the flat surface of a very large sheet of paper, with creatures living on it, like a small bug on the paper, constrained to forever travel *on* the surface but never *out* of it. The bug can move two dimensionally on the paper as shown. Now take the paper and wrap it around on a cylinder with radius R, continue to wrap it round and round (don't squish the bug!) until R becomes *much* smaller than the bug's size, too small for the bug to notice the extra dimension anymore. The bug now no longer lives on a 2-dimensional sheet-world, but rather lives on a 1-dimensional line-world and can *only* travel front and back like a tight-rope walker! Congratulations, you have just performed a Kaluza-Klein compactification.

Of course, it is a "little" bit more complicated to get rid of the six or seven extra dimensions that we are faced with in string/M-theory and it turns out that there are many different ways that this can be done involving quite exotic mathematics. Furthermore, once dimensionally reduced, it so happens that the theory *itself* is changed, new stuff appears with completely new behaviors. Disastrously, the number of possible ways one may dimensionally reduce those higher dimensional theories down to four spacetime dimensions allows for a truly large number of possible outcomes, each representing a completely new theory; so many, in fact, that they are collectively known as the *string theory landscape* [9, 10]. The often quoted estimate of the number of these product theories is the staggering 10^{500}! It is further assumed that the particular universe we live in arises as one of these possibilities. Well, if it is true that the physics we observe is just one of 10^{500} possibilities, what of the remaining ($10^{500} - 1$) solutions that do *not* describe our universe? Why are they there in the first place? Did Nature "choose" ours from this huge number randomly or is there a yet-unknown mechanism that picks particular universes from this vast number? Or, if we invoke the so-called *anthropic principle*, do all of the 10^{500} possibilities exist as separate universes each with a different set of laws of nature and we just happen to live in one of them (one that happens to be suited for intelligent

[5] The fact that the required number is *not* four is unimportant for the current argument and will be further discussed below.

life to evolve)? This whole set of questions is of course based on a series of *if* 's: *if* string theory is correct, *if* we understand compactification mechanisms correctly, *if* supersymmetry survives dimensional reduction (one of the criteria for choosing compactification mechanisms), *if* we can figure out the exact nonperturbative form of string theory, and so on and so forth.

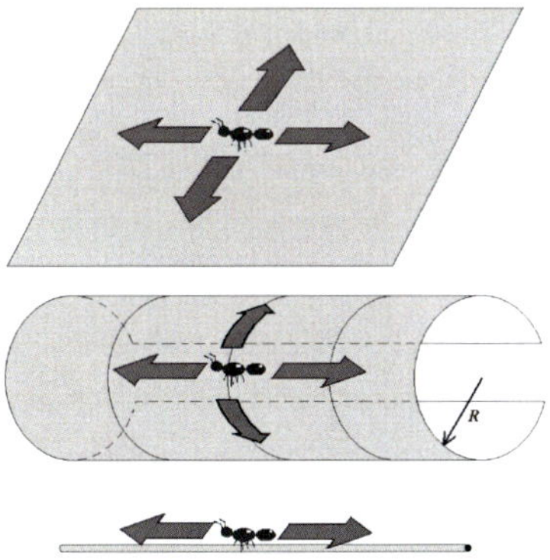

Figure 1: A bug living on a one-dimensional line world is unaware that there is a "hidden" extra dimension wrapped too small for it to notice. The arrows show the allowed directions of travel of the bug.

OH DEAR ...

Well, do we at least *know* what these product landscape theories look like? Do we have a way of generating them? Yes and no. While we can, in principle, set down the criteria that these theories must satisfy, the numbers and possibilities are of such sheer magnitude that such a task is, currently, quite hopeless. Some of these theories are quite known, however, and theorists have been exploring them for years, but still, compared to the huge number of possibilities that *is* the landscape, we are still only barely scratching the surface. The current situation with string theory is like that of a huge beautiful Persian rug that is being woven thread by thread. Normally we would start at the bottom and work our way systematically through the elaborate designs such that at any given time, we can see the completed portion of the rug all at once. Even if, as often happens via some flash of brilliance, the work starts at the center and proceeds outwards, the completed parts will still be overall solid, one thread leading smoothly to the next. Unfortunately in the case of string theory, the unfinished parts are not all in one place; they are scattered all over the rug, in Swiss-cheese fashion, or like a moth-eaten rug. It requires a considerable strain on the imagination to visualize what the finished rug would eventually look like. It has further been pointed out that given the huge number of possibilities; those parts of the rug that we *do* understand are, in fact, microscopically small, compared to the full landscape – see (Fig. **2**).

Figure 2: The string landscape Persian rug. Left: The way it looks right now, each color patch represents a landscape theory that is somewhat understood, and each thread represents a known mathematical connection to some other theory. This is *not* to scale; the number of known theories is *significantly* less than the unknowns. Right: Presumably it all represents something truly beautiful, *if* we can fill in the gaps.

THE BIG QUESTION IS THEN...

In my own research interests I have never dealt directly with strings, but rather with supergravity theories; long considered as low energy approximations to the full string theories. Approaching string theory in this way is actually quite common nowadays. Ever since the so-called second string revolution in the 1980's, interest in these approximations has soared, particularly because one can perform exact nonperturbative calculations in them, providing us with a glimpse as to what the full nonperturbative theory would be like, once it is found. This meant that more work is being done in the supergravity regime than on the superstrings themselves. For example, I am currently interested in one small part of the string theory landscape known as five dimensional N=2 supergravity. This theory is one of those that we know, before hand, to be totally unrelated to nature[6]. In other words, it belongs to the ($10^{500} - 1$) group of theories. Naturally I get asked: "Why then do you work on it? Shouldn't you, as a physicist, be interested in what describes nature? Why waste your time on something that you know *a priori* to be wrong?" Another closely related question is "What if, somehow, someone proves experimentally that there are no extra dimensions, or that supersymmetry does not exist or even more directly that subatomic particles cannot possibly be made of strings? In that case not only is the particular corner of the landscape you are working on wrong, the *whole* edifice has collapsed!

What will you do then? Will you drop your research and switch to something else? Or will you stubbornly continue to work on the (now wrong) string hypothesis? What will happen to all of your careers? And why take the risk in the first place?" These are of course very reasonable questions that may be rephrased as follows: "Are there any accomplishments of string theory that would survive such a total falsification?" It turns out that there are. Consider this: The lack of experimental results guiding us through the vast string landscape left string theorists with no choice but to systematically explore all of it[7]! These explorations, even within theories that we already know are not related to nature, have resulted in the discovery of deep and elegant mathematical results. Mathematicians today work in parallel with string theorists to explore the frontiers that the latter have opened. A famous example is the so-called *mirror symmetry* , discovered and developed by string theorists. This is not the first time of course that physicists have discovered mathematical theorems or even started entire branches of mathematics. In fact, let's not forget that calculus itself was discovered and developed by a physicist; namely Sir Isaac Newton [8].

Aside from advancing abstract mathematics, there is actually a possible practical application of string theory even if it turns out to not describe nature. An idea known as the AdS/CFT conjecture relates some string theories to other non-stringy ones via certain mathematical transformations [11]. The current hope is that this would lead to a way of solving hard (but practical) problems in the Standard Model by first addressing easier ones in the AdS/CFT-related string theories (which may themselves be non-physical). This, if it happens, would certainly be a major breakthrough, and may by itself guarantee the survival of string theory in some form, even if falsified by experiment.

Finally, studying the large number of theories in the landscape and how they are related to each other has provided, and continues to provide, deep insights on how a theory of physics generally works. While the Standard Model does describe nature, it is, at the end of the day, only a single theory. To understand it better, it is perhaps worthwhile to spend some time in the vast and rich laboratory of the string landscape where we can conduct mathematical experiments on a large number of specimens, further allowing us to understand the ones that we actually care about. It would be like a zoologist studying, say, the behavioral patterns of the African elephant by conducting experiments on a landscape of other species of animals to compare and contrast and further develop her understanding of the elephant.

So even if somehow, somewhere, someone shows that the universe cannot be based on string theory [9], I suspect that people will continue to work on it. Because it wouldn't be a valid description of nature, it may no longer be considered physics, nor will mathematicians, I suspect, accept it as pure mathematics, perhaps for not being abstract

---

[6] Partially because it predicts a large number of massless scalar fields; the so-called *hypermultiplet scalars*, something that is unobserved in our universe.

[7] Not that we are even *close* to scratching the surface.

[8] Leaving aside the famous historic dispute on precedence with the mathematician Leibnitz.

[9] One possible way for this to happen is the discovery of a new theory of everything that contradicts string theory but is backed by experiment.

enough. One can imagine that string theory in that case may define its *own* new discipline. A midpoint between physics and mathematics. A mathematical science that is devoted to the study of the structure of physical theory and the development of computational tools to be used in the "real world". Of course, if this happens, then one can expect that the number of string theorists or string theorist *wannabees* will drastically drop and the critics will be appeased. Instead, the theory will end up being studied by a healthy mixture of physicists and mathematicians who may no longer consider themselves either. They will continue to derive beautiful mathematical formulae and feed them to the mathematicians next door. They also might, every once in a while, point out interesting and important facts concerning the inner machinations of a physical theory that may serve to guide the physicists exploring the actual theory of everything over in the next building. Whether or not string theory describes nature, there is no doubt that humanity has stumbled upon an exceptionally huge and elegant structure that might prove to be very difficult to abandon. If disproved as a theory of physics, it may simply become its own field. The formation of a new science or discipline is something that happens all the time. For example, most statisticians nowadays do not consider themselves mathematicians anymore. In fact, in many an academic institution, departments of mathematics now call themselves "mathematics *and* statistics". Some have even already detached into separate "departments of statistics". Perhaps the future holds a similar fate for the "unphysical" as well as "not-so-purely-mathematical" new science of string theory.

REFERENCES

[1] B. Greene, *The elegant universe.* Norton, 1999, p448.
[2] L. Smolin, *The trouble with physics.* Houghton Mifflin, 2006, p392.
[3] L. Smolin, *Three roads to quantum gravity.* Weidenfeld & Nicolson, 2000, p 231.
[4] K. Popper. *Conjectures and Refutations.* Routledge and Keagan Paul, 1963, pp33-39.
[5] P. Woit, *Not Even Wrong,* Cape, 2006.
[6] P. Woit, "Is string theory even wrong?" *Sci. Am.* 90, 110, 2002.
[7] P. Woit. "String Theory: An Evaluation", *http://arxiv.org/abs/physics/0102051.*
[8] P. Woit, *http://www.math.columbia.edu/~woit/wordpress/*
[9] T. Banks, M. Dine and E. Gorbatov, "Is there a string theory landscape?" *J. of HEP*, 0408, 058, 2004.
[10] R. Bousso and J. Polchinski, "The string theory landscape", *Sci. Am.* 291, 60, 2004.
[11] J. Maldacena, *Large N field theories, string theory and gravity.* ICTP Spring School on Superstrings and Related Matters, Trieste, Italy, p2-10, Apr 2001.

CHAPTER 4

String Theory in the Classroom. A Case Study

D.H.J. Cho

Department of Physics, Kenyon College, U.S.A

Abstract: The first question every instructor asks herself when she teaches a class for the very first time, is what topics should be included and what topics should be discarded. This is usually a tricky balancing act relying not only on conventional wisdom, but also on personal taste and judgment. In this short article, I describe my experience including string theory into a non-science major general education class called "Origins", offered at Kenyon College during the fall semester of the academic year 2009-2010.

INTRODUCTION

2010, I co-taught, for the first time, a semester-long class titled "Origins". The class covered the origin of the universe among other topics. Previous instructors of the course focused mostly on standard textbook treatments of the Big Bang theory with a very brief mention of the inflationary scenario at the end. I had decided to reverse the focus to the inflationary scenario with the standard Big Bang model as a motivating background. It was a rather unusual choice. I thought, however, that the inflationary model was a more appropriate topic for a course specifically designed to discuss the origin of the universe. Moreover, I believed that the model had enough observational evidence to be considered as a standard paradigm for solutions to problems that marred the standard Big Bang scenario.

I went further, however, and made an even more unusual (mildly put) choice of including string cosmology and loop quantum cosmology. These are the two main contenders if we were to ask even more fundamental questions the inflationary model fails to address. Namely, what started inflation? what happened before inflation? And how did the universe really begin? These are very interesting and grand questions not only in science, but also in philosophy. Many people, including my students, are interested in these questions, and string theory and loop quantum gravity can provide possible answers to them. Now, one "tiny" problem was that the class is offered in the physics department as a science class, but can these models really be considered science? Without a single shred of experimental evidence, they are not part of science in any normal sense. Then how could I justify myself in teaching them in a science class? I attempt to describe my reasoning in the rest of the article, after a short description of the class itself.

COURSE STRUCTURE

The class was divided into two halves. Roughly, the first half, my part, was about the origin of the universe as a whole. It was then followed by a discussion of the geological and biological origins on Earth. My part was again subdivided into three parts: Discussions of the standard Big Bang model, the inflationary model, and current frontiers. The main text for the first two parts was "The Inflationary Universe" by Alan Guth [1], an excellent description of both topics at an accessible level. The last part covered various ideas on the beginning of the universe. The topics were string theory, the string landscape, loop quantum gravity, loop quantum cosmology, pre-Big Bang models, the ekpyrotic universe model and the dynamical triangulation approach to quantum gravity[1]. For rather obvious reasons, I chose not to use a single textbook (No such book exists as far as I know) and rather rely on six different articles from Scientific American [2-7].

The main focus in the first two parts and the last part was also slightly different. While the former included a lot of detail (as much as is naturally possible at this level), scientific ideas and observational evidence, the latter emphasized big questions and big ideas. For example, in the first two parts the students learned the physics of the

[1] *Editor's note*: Most of these topics will be discussed later in the book in more detail.

Doppler Effect and how it relates to Edwin Hubble 's discovery of the expansion of the universe, of the physics of recombination and how it relates to the Cosmic Microwave Background Radiation (CMBR), of phase transition and how it relates to inflation among others. On the other hand, the last part focused more on a brief description of the main aspects of quantum gravity models and their claims.

Since this is an article on string theory, I will focus on the string theory part of the course, and describe it in a little bit more detail. To motivate the question, "why strings?" I started by very briefly describing the historical background of string theory as a dual resonance model for the strong interactions. They have already learned about quarks, gluons and asymptotic freedom earlier in the class. Once students understood the phenomenology of asymptotic freedom, it was quite easy to argue for a string-like object between quarks. String phenomenology, especially the unification of forces, was next. Here, I demonstrated standing waves and introduced the concept of normal modes; musical harmonics and all that. Higher dimensions and the Kaluza-Klein mechanism were much more challenging, however. Pictures of lower-dimensional objects to explain some geometric ideas helped, but not enough. It was quite hard for the students to distinguish between the intrinsic and extrinsic properties of geometry. They automatically assume that everything is embedded in three dimensional space! This problem is not, however, a unique problem in string theory. The problem already arose when I taught the expanding universe: expanding into *what*, they asked. An even tougher problem was to explain the Kaluza-Klein mechanism. There are two conceptual issues here. One is a scale issue and the other is the concept of a compact space without a boundary. Scale dependence and effective descriptions of physics were taught with an analogy of ants and a bird on a wire. Compact space with no boundary (it's easy in two dimension once students overcome the earlier problem with intrinsic geometry. A sphere is a perfect example) was taught using the example of a pacman universe, which is effectively a 3-dimensional torus. The final hour of the string part was devoted to a description of the string landscape. This was perhaps the most controversial part as long as a choice of topic was concerned, and will be described in more detail in a separate section below.

PROS AND CONS

Considering what to include in a course is a tricky balancing act. The instructor should carefully weigh the pros and cons of each topic to be taught under the given time constraints. Each course in a curriculum has a fairly specific set of goals. Therefore, in any course, there are core material that has to be included. It is the left-over gaps that are filled with more personal choices of topics. String theory is clearly not core material, even for a class like "Origins". The question then is can it be a gap-filler? Here are some of the pro and cons I weighed when I made the choice.

One of the pros is student interest. At least partly due to media reports and a large number of popular books, string theory is very popular among the current generation of students. "String theory" they think, however, is not necessarily the string theory in any textbook sense. My experience from talking to students taught me that "string theory" means a much broader concept that represents speculative, crazy, fundamental, weird, wacky, very difficult and nerdy ideas about the beginning of the universe and the fundamental atoms of space and time.

Students tend to get excited about wacky and strange ideas they heard about but don't understand. The air of mystery surrounding very strange ideas attracts students. I still vividly remember the first time I read about black holes, spacetime foams, and string theory ; the very idea of the unification of physics. I was in middle school and had just begun learning physics. Naturally, I really didn't understand any of these concepts, but they sure sounded mysterious, wonderful and fascinating. I was hooked. If being interesting and weird are the only criteria, however, there are tons of other wacky ideas about the origin of the universe and the fundamental laws of physics. What distinguishes string theory from those others? In my mind at least string theory is a legitimate development from firmer, more orthodox, branches of physics (Along with a few other theories like loop quantum gravity). It includes gravitons, the Kaluza-Klein idea, the full edifice of modern quantum gauge theory, and much more.

Now, here are some of the cons and my excuses for ignoring them: I mentioned earlier that string theory is worth inclusion in the course, but at what cost? Obviously there are many more topics than can be fit in the available time, so why string theory? Why not teach them more about, say, the galactic structure of the universe, the so-called CP problem in physics or Baryogenesis, all being legitimate, non-too-speculative issues in orthodox physics? I don't think I have a really convincingly strong argument here. Ultimately, it was a personal choice. It seemed to me that string theory would be a more fun topic to be included over, say, Baryogenesis (To be clear, I did cover Baryogenesis and structure formation, but not in detail).

In my opinion, a more serious problem is whether string theory teaches us any new science? When I design a course like "Origins", I try to emphasize the physics background of the models and ideas under discussion. For example, the expansion of the universe is closely tied to the physics of the Doppler Effect, the distance ladder and Friedmann 's model of the universe. That is, teaching the expansion of the universe is a good chance to teach other important physics and astronomy concepts that are used in other areas too. Does string theory have anything unique to teach students? Not really, except possibly the idea of higher dimensional space which is not even an experimentally verified fact. So, what is the value of teaching string theory?

I don't think that there is any specific scientific knowledge unique to string theory which can be taught to students at this level. In my opinion, however, the educational benefits of it come from an entirely different place. A course like "Origins" is typically taught at the level of a general audience. One of the goals of these liberal arts courses is to nurture student creativity. This doesn't just come from knowing how things work. It comes from knowing various different ideas, even crazy ones, in many different disciplines. String theory may not have a specific *science* to teach yet, but it is a fruit of the general physics paradigms of reductionism, mathematical esthetics as a criterion of the theories of nature and the unification of the laws of physics. The inclusion of string theory in the course, therefore, should be understood as teaching the cumulative realization of the philosophical ideal of (high energy) physicists.

Finally, the biggest reservation I had was with the string landscape. The question is a hotly debated one. Is the string landscape even a science? To make a long story short, I think that string theory itself *is* a science, but the string landscape is a very grey and risky borderline.

DOES STRING THEORY BELONG TO SCIENCE CLASSROOMS?

Is string theory a science? In the strictest sense, it isn't. There is not a single direct experimental evidence for it. Moreover, we may not be able to have a direct detection of the strings themselves in the foreseeable future simply because the energy scales associated with it are too large. Strictly speaking, therefore, it is not a scientific theory. Furthermore, there is no general formalism of the theory that everyone agrees upon. At best, string theory is a collection of educated guesses based on firm scientific ideas. So, does string theory belong to a science class? I think so. In a bit of twisted sense, I even think that it is a great topic.

Let's make a zoological analogy. Typically, a zoology class discusses the animals found in nature, their classifications and characteristics. Learning about them is interesting, but even more interesting is learning about an animal that we don't know much about yet. Suppose we have an animal that we could not capture to study in a controlled manner. Rather, this is an animal that has only been sporadically spotted, and observations about it are scarce and seemingly contradictory, for example the Loch Ness monster or Bigfoot. It is not possible to discuss such a creature coherently. We may not even be able to tell what kind of a beast it is, or to what branch of the evolutionary tree it belongs. We can, however, learn about whatever clues and guesses about it that are available. The majority, if not all, of these ideas will eventually turn out to be false. But, does it really matter?

In most science classes we teach firmly established scientific knowledge, but we rarely show the scientific process that established them in the first place. We don't, for example, emphasize the great debates that surrounded the early development of quantum mechanics. We may perhaps spend a couple of lectures on ancient Greek astronomy and the Copernican revolution as examples of the scientific process. These are all great examples no doubt, but there is an argument to be made about the discussion of contemporary topics, however controversial they may be. Students are excited by the fact that their teacher does not know the answer (indeed, no one does). It is a new frontier that they can think about and claim as their territory unlike most of the subjects they learn in the standard classes. That is exactly where the merit of teaching string theory at this level lies. With carefully designed homework sets and guidance from the instructor, such a class can be a live stage for learning the scientific process: Clues, motivations, ideas and the fact that most of these ideas will turn out to be wrong, not to mention the hot debates surrounding the legitimacy of string theory being a scientific theory in the first place.

I should, however, point out one lingering worry I had regarding the string landscape. String theory predicts a large number of possible universes and cannot, in principle at least within string theory itself, tell us which of these is our universe. One natural remedy is by invoking the anthropic principle. Proponents of the landscape scenario argue that the landscape is a generic feature of string theory. First of all, however, I am not sure whether we understand string theory enough to conclude that the landscape is a generic feature of it. However, maybe we are just missing something and a future theory (whatever that may be) will not have that feature. The details of the issue are also not that simple and there are efforts to extract predictions out of the landscape. What I am worried about, however, is that we cannot teach those subtleties at this level, and a naïve string landscape seems awfully close to being "not even wrong"[2]. In a sense, a statement from the string landscape scenario (possibly) implies an infinite number of universes that even in principle[3] cannot be detected, but still exist anyhow. This statement is not a scientific one since it cannot be tested experimentally. Teaching a model that contains non-scientific statements as a scientific model is a dangerous act, especially in the era of Intelligent Design and similar pseudo-religious theories that try to pass for science.

CONCLUSION

Based on my experience outlined here, I believe that teaching string theory as part of a general science class has merits, probably not so much as scientific knowledge, but rather as an example of the scientific process. The focus of such a class, therefore, should be on ideas, motivations and debates rather than detailed descriptions of a given theory. An instructor should also be careful to point out controversies such as the one surrounding the string landscape and its scientific legitimacy.

APPENDIX: THE POSSIBLE VALUE OF STRING THEORY IN AN UPPER LEVEL PHYSICS CLASS

In this article, I exclusively describe and argue string theory as a topic in a general science class. How about the value of a serious string theory class as an upper level physics course? There is at least one textbook that is based on its author's upper level undergraduate course at MIT [8]. Again, the central question to me is: at what cost? In my opinion, a class in string theory cannot offer much in terms of new physics at the undergraduate level, at least in most schools. Bosonic string theory at a basic level is a theory of the quantized one dimensional string. I doubt that students can get much beyond a one dimensional quantum field theory. If that is the case, I think it is better to teach students a proper scalar field theory rather than a string theory or some other controversial topic.

REFERENCES

[1] A. Guth, *The Inflationary Universe*. Basic Books, 1998.
[2] R. Bousso and J. Polchinski, "The string theory landscape", *Sci. Am.,* pp79-87, September 2004.
[3] J. Ambjørn, J. Jurkiewicz and R. Loll, "A self organizing quantum", *Sci. Am.*, 299N1, pp42-49, 2008.
[4] M. Bojowald, "Following the bouncing universe", *Sci. Am,* pp44-51, October 2008.
[5] L. Smolin, "Atoms of space and time", *Sci. Am.,* pp66-75, January 2004.
[6] M. Tegmark, "Parallel universes", *Sci. Am.,* pp40-51, May 2003.
[7] G. Veneziano, "The myth of the beginning of time", *Sci. Am.,* pp54-65, May 2004.
[8] B. Zwiebach, *A First Course in String Theory*. Cambridge University Press, 2009.

[2] *Editor's note*: See footnote 7 on page 20.
[3] There are attempts to see if we can test the existence of other universes. One idea is that if there are other universes, they may have collided with ours in an early epoch and may have left imprints somehow on, say, the cosmic microwave background.

CHAPTER 5

Science Fiction of Everything

Florian Conrady

Perimeter Institute for Theoretical Physics, Waterloo, Ontario, Canada; E-mail: fconrady@perimeterinstitute.ca

Abstract: In this article, I will describe my perspective on the present state of fundamental theoretical physics and my guess on what will be the key elements of future theories of everything. First I will review three important approaches: string theory, loop quantum gravity and noncommutative geometry. I will explain their main ideas and strengths, but also point out what they are still missing on the road to providing a true unification of physics. Based on this, I will speculate about how a future theory of everything could look like: I expect that its basic degrees of freedom will be networks or matrices. Its principal achievements will be the unification of spacetime and fields, and the dynamical generation of spacetime. I will also argue that this scenario opens the way to many interesting conceptual possibilities: the existence of several spacetimes or realities that interact with each other; that spacetime and the laws of nature are not "God-given", but evolved and contingent; that the very structure of reality could exhibit higher levels of organization. The thread that runs through the discussion will be renormalization : I will start by explaining nonrenormalizability and conclude with remarks on nonperturbative renormalization.

INTRODUCTION

The main argument of this article is that spacetime has to be unified with the other concepts of physics. A true theory of everything must explain all of reality in terms of one principle. This should include spacetime and not let it stand out as a notion that is completely different from the rest.

The three principal players that appear in current theories are forces, matter and spacetime. Forces are described by fields such as the well-known electromagnetic field. In addition, there are the weak, strong and gravitational forces[1]. Matter particles like electrons and quarks correspond to fields as well, although they are of a different type than the force fields. Finally, there is spacetime: the container or stage in which forces and matter evolve and all physical processes take place. The gravitational field is special in the sense that it characterizes not only gravity, but also the geometry of spacetime. Thus, it can be seen as a *property* of spacetime. Due to this, the word spacetime is sometimes used synonymously with the gravitational field.

To understand this article correctly one should keep in mind, however, that spacetime and the gravitational field are *not* the same. Spacetime is much more than geometry and gravity ; it also has another, more elementary property known as *topology*. The term originates from the Greek "topos"', which means "place". Put simply, topology is the information about how different parts of spacetime connect with each other. It tells us, for example, whether these parts form a line or a circle, a disk or the surface of a sphere, or more complicated structures. Geometry, on the other hand, is about lengths, sizes and angles: when the topology is such that two regions of spacetime are connected with each other, geometry tells us how far apart they are. It determines the length of any line that goes from one region to the other. In the previous examples, geometry is about the size and radius of the circle, the size of the disk or the radius of the sphere.

I emphasize the difference between geometry and topology, since it is here that a divide occurs in today's established theories of physics. We have come to learn that the geometry of spacetime is dynamical and variable similarly to the strong and electroweak fields. In Einstein 's theory of general relativity there are equations that govern the evolution of gravity and geometry. While there are conceptual differences, gravity and the other forces are clearly related.

In contrast, the more basic part of spacetime that is not geometry, i.e. topology, is kept apart. There are no equations for it. Instead it is postulated at the outset and seen as immutable. The way in which different points of spacetime

[1] The electromagnetic and weak forces can be understood as consequences of a single force known as the electroweak force.

Moataz H. Emam (Ed)

connect with each other cannot be changed. Moreover, as a mathematical concept, spacetime is very different from the notion of fields. Fields reside *in* spacetime, but they are *not* spacetime. Simply put, spacetime is a set of points, while fields assign values to these points. As a result, we have a situation where the force and matter fields are on one side of the divide, and spacetime is on the other.

In this article, I will review three lines of research that go beyond experimentally verified theories of physics: string theory, loop quantum gravity and noncommutative geometry. In doing so I will ask, in particular, if they are able to provide a unifying principle for all of physics, and if they can bring together the two sides of the divide I just described. We will see that none of these theories achieves this unification in the full sense. Moreover, the separate status of spacetime is related to some of the unresolved problems in these approaches.

Based on this, I will argue that a theory is needed in which spacetime is treated on the same footing as force and matter fields. In particular, spacetime topology has to be variable like the fields and subject to equations. I will speculate about future theories of everything that could achieve this feat and how they might look like. My expectation is that *labeled networks* or matrices are suitable variables for this purpose. It will be explained why networks and matrices are closely related and that they could be universal enough to encompass the concept of field and spacetime. In the section titled "Science Fiction" I will carry speculation even further and think about the fascinating conceptual implications of a theory in which spacetime and fields are one.

What I promote in this article is in a sense a continuation of relativity. Special relativity taught us that space and time are relative, and replaced them both by the concept of an absolute spacetime geometry. Then, general relativity showed that this geometry is not absolute, but rather dynamic, and the only entity that remained fixed was bare spacetime: i.e. spacetime without geometry. The logical continuation of this is to make relative what is left: spacetime topology or the way points of spacetime are connected with each other.

I should emphasize that this article is speculative. There is currently no experimental evidence that string theory, loop quantum gravity or noncommutative geometry are on the right track. The same holds true for the hypotheses I add to these theories. Therefore, I do not want to convey the impression that "theoretical physicists have shown that X is true", where X is one of the speculations in this article. We are talking about possibilities here, and what I will discuss may very well be wrong.

The article is organized as follows: I will start by explaining nonrenormalizability in quantum field theories, since it is a primary motivation for string theory. This section will also serve to introduce some concepts that will frequently appear in the rest of the article. Sections III, IV and V give a brief review of string theory, loop quantum gravity and noncommutative geometry respectively. Section VI describes my own speculations about future theories of everything in which spacetime and the rest of physics will be unified. This is followed by section VII, where I will discuss the potential conceptual and philosophical consequences of such a scenario. In the last section, I will come back to the topic of renormalization and conclude with remarks about nonperturbative renormalization.

Since this article is intended for a general audience, no references to scientific papers are given. Nor will I mention the names of the people who have discovered the things I talk about. If I did, I would have to do it fairly, and then I would have to mention so many names that it would inflate the text unduly. I will only list four references that provide the interested reader with basic technical introductions to some of the topics [1-4].

NONRENORMALIZABILITY IN QUANTUM FIELD THEORY

A principal motivation for string theory is the nonrenormalizability of perturbative quantum gravity. To understand this statement we need two concepts: *renormalization* and *perturbation theory*. Renormalization has to do with the fact that theories appear different depending on the scales we look at.

Suppose we have a fundamental theory defined at very short distance scales. If it is a *field theory*, its variable is a *field*. A *field* is a function that assigns a number (or several numbers) to each point in space and time. The name is motivated by the example of electromagnetism, where an electric or a magnetic field corresponds to an assignment of a field's strength and direction to each point in space. The effects of this field can be seen, for example, when iron filings are sprinkled around a magnet.

The field is a variable, since it can, in principle, take on many values. Every way of attributing field values to all points in spacetime is a possible choice. I say "in principle" since there is another part of the theory – its equations – that tell us that some fields are better than others. Some choices of the field are solutions to the equations, and some are not. These equations are the natural laws of the theory and they specify which choices of the field are in accordance with nature, and which are not. This also allows us to determine what the theory predicts. Since a solution satisfies the equations, its values are not arbitrary: they will obey certain relations. For example, the field values at earlier times will be related to those at later times. In this manner, the theory tells us how the field will evolve in the future, if we know some values of the field in the past. It will also tell us how the field values influence each other in the course of such an evolution.

There is a general scheme for determining what the equations of a theory are. The central quantity for this is a function of the fields called the *action*, usually denoted by the letter S. It would be too mathematical to explain all the details, so let me just say that there is an algorithm that starts from the function S and produces the equations for the field as an output. Since the action S fixes the equations uniquely, it can be used as another way of specifying the natural laws of the theory – see the diagram in (Fig. **1**).

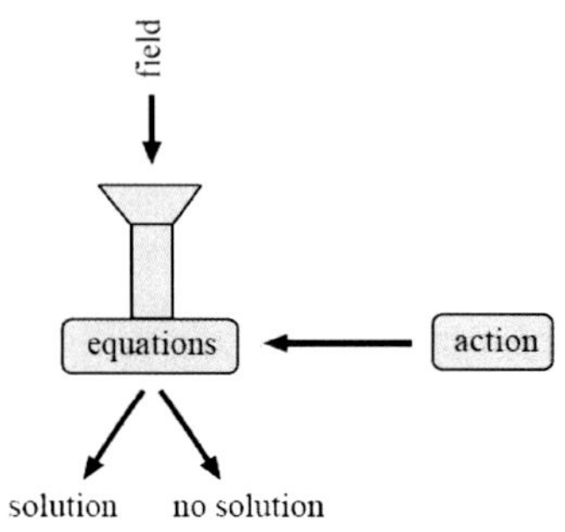

Figure 1: The action function determines the equations of a field theory. The equations, in turn, specify which fields are solutions and which are not.

The action is a function of the field, and hence a function of the many field values at different points in spacetime. It combines these values by adding them, by multiplying them with each other or with constants, and by various other means. Typically, the action function is a sum of a number of terms, where each term is multiplied by a numerical constant (say c_1, c_2, c_3 etc). These terms have a physical meaning, since they affect the form of the equations and their solutions: they represent the different types of interactions that occur between field values. The constants in front of the terms describe the strength of each interaction and represent the natural constants of the theory. They are also referred to as *coupling constants,* since they specify how field values "couple" to (i.e. interact with) each other. An example of this would be electrodynamics and the interaction between the electric field and the matter field of an electron. This interaction arises from a particular term in the action. Its strength is given by the electric charge of the electron and the latter appears as a constant in front of that term.

An important property of field theories is that they predict the existence of particles. When one studies the solutions of a field theory and how they behave in time, it becomes apparent that they automatically contain particle-like objects. Very roughly speaking, the particles manifest themselves as particular patterns in the field that move through space. This is the reason why field theories, or more precisely, *quantum* field theories, are so important in elementary particle physics. All the fundamental theories that have been experimentally confirmed so far are quantum field theories.

When we think in terms of particles, the terms in the action correspond to different types of interactions between particles, and the constant in front of each term determines the strength of the interaction. This is just another way of saying what I stated before using the language of fields. I repeat this point, since it will play an important role in the rest of the article.

For practical purposes, it is very important to be able to zoom out from microscopic distance scales and to know how a system behaves at much coarser distance scales. This is analogous to reducing a picture of high resolution to one with lower resolution. In general, the microscopic theory with action S will be described at larger scales by a new action S' that is different from S. S' is called the *renormalized* action that follows from the fundamental action S when we zoom out to large scales.

As I said, the constants of nature appear as constants in the definition of the fundamental action S. In general, these "constants" change as we go from small to large distance scales. That is, the constants in S are different from the ones in the renormalized action S'. Usually, it will be the constants in S' that we experimentally measure as particle masses, charges or other quantities, not the ones in S. The reason for this is that our experiments can only achieve a limited resolution that lies far below the resolution needed to access the microscopic scale of the fundamental theory. We can therefore only observe how the theory behaves at coarse scales, as described by S' and its *renormalized* constants. The original parameters and constants of S are not directly accessible. The theory of *renormalization* provides us with equations that determine S' as a function of S and therefore the relation between the unobservable microscopic constants and the ones measured experimentally. These are called the renormalization group equations. We learn from them, in particular, that only certain values (or ranges of values) of the microscopic constants are consistent with the observed constants of nature. If one chooses, for example, some random value for the microscopic mass of the electron, it will most likely lead to a large scale electron mass that differs from the one measured in experiments. Hence one has to tune the microscopic mass suitably, so that the resulting renormalized mass agrees with observations. Another important aspect of renormalization is that a given value of a microscopic constant c does not only influence its own renormalized constant c', but can also affect the renormalized values of *other* natural constants. Thus, a constant c_1 can cause another constant c_2 to grow to a value c'_2 in the renormalized action, even if c_2 was initially zero. In this case, one says that the renormalized constant c'_2 has been *generated* or *switched on* by renormalization. In regards to this generation process one has to distinguish two principal types of natural constants: These are called *renormalizable* (or *relevant*) and *non-renormalizable* (or *irrelevant*). The first kind has the property that it generates only a finite number of new renormalized constants at larger scales. Such physical constants are relatively easy to deal with in calculations and this is the reason why they are named renormalizable. They are also called relevant, since they have the tendency to grow larger under renormalization. That is, the strength of these constants increases as one zooms out to larger scales and thus they are important in the description of large-scale phenomena.

In contrast, the second category of constants creates an *infinite* number of new constants under renormalization. This, more complex, behavior poses considerable conceptual and technical challenges. In fact, it was thought initially that such constants and their associated interactions cannot be treated by renormalization and hence they were labeled nonrenormalizable[2]. Furthermore, these constants behave contrary to the renormalizable ones in the sense that they grow weaker at large scales. Therefore, they only have measurable effects at small scales (high energies) and become irrelevant in large-scale (low energy) processes.

The second concept we need is *perturbation theory*. Generally, the dynamics of theories is very complicated and we often only understand a small number of solutions of a theory, while the entire set of solutions is much bigger. This is particularly true for gravity. This problem can be somewhat alleviated by applying a method called perturbation theory: it allows one to obtain additional approximate solutions by slightly perturbing a known exact solution. This approximation works as long as the difference between perturbed solution and exact solution is sufficiently small. Therefore, perturbation theory can only give us new solutions that are very close to the original solution.

In many theories, our knowledge is restricted to these islands of perturbations around exact solutions. In some cases, the theory is not even *defined* outside the realm of perturbation theory. That is, the construction of the theory itself is based on perturbation theory. In this case, one speaks of a *perturbative* theory. Examples of this are the theory of electroweak interactions and perturbative string theory.

Many questions in elementary particle physics can be answered very successfully by combining the techniques of perturbation theory and renormalization. Perturbation theory gives us knowledge of physical processes that are close to certain exactly known solutions of the theory. And renormalization tells us what the theory predicts when we zoom out from the smallest scales to the scales that are accessible to experiments.

---

[2] From the modern perspective, this nomenclature is somewhat confusing, since we now know that renormalization can be applied to nonrenormalizable interactions.

However, when it comes to quantum gravity this scheme encounters difficulties. The property that is responsible for these problems is called *non-renormalizability* and has to do with the non-renormalizable natural constants mentioned earlier. Let me explain what happens. In the case of gravity the variable quantity is the geometry of spacetime itself. The action S is a function of this geometry and consists of a few terms that involve the curvature of spacetime. This action encodes the laws of general relativity without matter.

With the help of perturbation theory and renormalization one can attempt to construct a quantum theory from this. As described before, perturbation theory tells us how the theory behaves near exact solutions, and renormalization allows us to get a large-scale action S' from the fundamental microscopic action S. In the context of gravity, the exact solution is called a *background spacetime,* since it plays the role of a fixed stage on which the evolution of the perturbed solutions takes place. The background can be likened to the unperturbed surface of a pond, and the perturbed solutions correspond to small waves traveling on it.

The problem with this approach to quantum gravity becomes apparent when S' is computed. It turns out that the natural constant of gravity; Newton's constant, is nonrenormalizable. As a result, renormalization generates an infinite number of terms in the renormalized action S', each equipped with its own constant. This is, on the one hand, a technical challenge, since it is very hard to do computations with an infinity of different kinds of interactions. On the other hand, it leads to a conceptual puzzle: Since the appearance of these terms is inevitable, there is no reason why we should not include them from the very beginning in the fundamental action S. Moreover, whether they are zero in S or not, we have no way of deducing the values of these microscopic constants, since they are irrelevant (in the sense defined earlier) and of no consequence to large scale physics. Thus, quantum gravity ends up containing an infinite list of possible interactions and associated natural constants c_1, c_2, c_3 *etc.* whose values remain unspecified.

Theories like gravity that have nonrenormalizable interactions are called *perturbatively non-renormalizable* and they all share these problematic features. In contrast, a theory is said to be *perturbatively renormalizable* when it has only renormalizable natural constants. In such a theory, renormalization operates only on a finite set of coupling constants and a finite number of measurements is sufficient to determine these constants. An example of this is the Standard Model of the electroweak and strong interactions.

The fact that we cannot avoid infinitely many terms in perturbative quantum gravity highlights a profound problem in quantum field theory: in the absence of any deeper principle the infinitely many constants are not determined by any theoretical considerations. They could have any set of values as long as they are consistent with the experiments available to us. The only way to fix them all would be to carry out an infinite number of experiments at smaller and smaller scales from which one could infer the value of each constant c_i. In a sense, such a theory is more of a template of a theory, since many of its features are not predicted theoretically, but have to be filled in by experiments. This is called the problem of *lack of predictivity*. One could compare quantum field theory to an irrational number with an infinite number of digits after the decimal point. Unless some rule is given that specifies all of the digits, it is not clear what the number really is. We could demand, for example, that the square of the number is 2, and this would fix all of the decimals. The analogue of such a requirement is missing for perturbative quantum field theory.

This should give a rough idea of what the problem of nonrenormalizability in perturbative quantum gravity is. At this point string theory comes into the game.

STRING THEORY

String theory has its origin in the following question: what happens if we replace particles, which are 0-dimensional objects, by 1-dimensional strings? At the end of the 60's people began to explore this possibility and it turned out that one can write down theories that have strings as elementary degrees of freedom.

The action S of string theory has a surface as its variable. This surface is called the *worldsheet* and it is a surface that is situated in spacetime. It can have different topologies (e.g. be spherical, donut-shaped or more complicated shapes), and there are many ways of arranging and positioning it in spacetime. Each choice of position amounts to a particular value of the variable. The action is a simple function of the worldsheet and is related to its area. It contains a constant that can be interpreted as the tension of the worldsheet.

What does all this have to do with strings? The worldsheet is the history of a string. A given moment in time corresponds to a surface in spacetime that can intersect with the worldsheet. The set of intersection points consists of zero, one or several lines or strings. If we consider a whole sequence of times t_i, each of the associated intersections can be regarded as a snapshot of the strings and the sequence shows how they move as time passes. In this sense, the position of the worldsheet is equivalent to the motion of the strings. Interactions of the strings are represented by the shape of the worldsheet. The worldsheet in (Fig. **2**) shows, for example, the splitting and merging of strings.

There are two remarkable facts that make the use of strings very interesting. Firstly, string theories include types of string motion that behave similar to known particles: to photons and gluons, which mediate the electroweak and strong forces, and to gravitons, which are expected from the hypothetical theory of quantum gravity. Thus, string theory has the potential of describing the electroweak, strong and gravitational forces in a unified manner, starting from only one principle! The second remarkable aspect is that the issue of renormalization is less problematic than in point-particle quantum field theories.

To explain these statements in more detail, I need to say more about the relation between strings and particles. The tension of the strings is chosen to be very large, so that a string can be stretched only very little. The string's typical size is of the order of the so-called Planck length l_p=10^{-34} centimeters. If one zooms out by a few orders of magnitudes these tiny strings will look very much like particles. For this reason, every string theory can be approximated by a quantum field theory, the so-called *effective field theory*: the particles of this field theory correspond to the strings, and the interactions of the particles reflect the various ways in which strings can merge and split. Different species of particles appear, because strings can have different ways of vibrating around their centers of mass, and from a large distance they look like particles with different masses, charges etc.

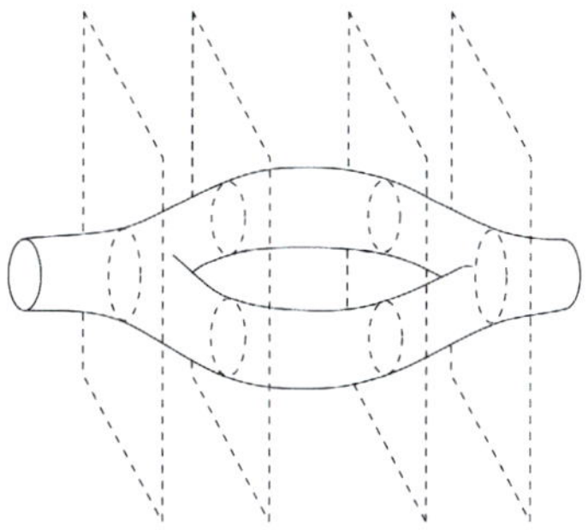

Figure 2: A string breaks into two strings which then merge again into one string. The horizontal direction corresponds to time.

We said in the previous section that the constants in a microscopic action *S* have to be suitably tuned, so that the constants in the large-scale action *S'* do not take on unrealistically large values. Moreover, in perturbative quantum gravity, this tuning even requires an infinite number of terms in the microscopic action *S*.

How is this situation improved in string theory ? The string action itself does not require the addition of new terms in order to have an acceptable physical behavior. The action of the associated effective field theory *does* contain an infinite number of terms, but they are all fixed by the choice of the string action! That is, the infinite number of different coupling constants can, in principle, be determined from the way the strings interact with each other. In this way, string theories solve the problem of nonrenormalizability. In the analogy between quantum field theories and irrational numbers, string theory plays the role of a number like e, π or $\sqrt{2}$ which specifies all of the infinitely many digits at once!

However, the achievements of string theory come at a price. It turns out that string theories have to satisfy certain constraints, since otherwise they become unstable, break symmetries that they should not break, or become too complicated to work with. One of these constraints says that the number of dimensions of spacetime in which the strings move has to be 10. Another constraint demands that string theories exhibit a symmetry under exchange of bosons and fermions[3]. This symmetry goes under the name of *supersymmetry*.

[3] Particles can be grouped into two main species with different properties: *bosons*, which are responsible for the forces of nature, and *fermions* which constitute matter.

These restrictions are a challenge for the theory, since the observed spacetime dimensions are 4 (3 of space and 1 of time), not 10, and the currently known particles do not exhibit any supersymmetry. To solve these problems, the use of the techniques of *Kaluza-Klein compactification* and *supersymmetry breaking* was proposed. It is postulated that 6 of the 10 dimensions of string theory are "compactified". This means that in 6 directions space is periodic and small, similar to a circle with a small radius. Due to their small sizes, these extra dimensions would not be observable with present experiments, and hence one would effectively see only 4 spacetime dimensions. Supersymmetry breaking borrows ideas about symmetry breaking from particle physics and assumes a scenario where, in a certain sense, supersymmetry is broken by the vacuum. As a result it looks *as if* supersymmetry was absent at large scales, although it is actually there.

When the requirements of symmetry, stability and simplicity are taken into account, one finds that there are five different string theories that pass all criteria. They are called type I, type IIA, type IIB, heterotic $SO(32)$ and heterotic $E_8 \times E_8$ superstring theories. Later it was discovered that these string theories are actually not independent of each other. They are related by the so-called *duality transformations* or simply *dualities*.

A duality is essentially a transformation that maps objects from a theory X into *similar* objects belonging to another theory Y. Like a rotation of an image, the transformation turns objects into *different* objects, but like the rotation it also leaves some of their properties unchanged. There are similarities between object and transformed object that allow us to say that features of the theory X are reflected in features of the theory Y. More precisely, if there is a formula for computing a physical quantity in theory X, there will be a formula for computing the same quantity in theory Y. The latter formula may look very different, and it may involve very different objects, but it still produces the same quantity.

Such duality transformations are fascinating, since they suggest that two different theories can just be different ways of representing the same thing. If this is true, the duality transformations are not so much transformations from one theory to another, but changes in the perspective from which one single theory is seen. In the analogy with rotations, this would correspond to the fact that the image of an object is different depending on how one rotates the viewpoint of the camera, but the underlying object is always the same. In the context of string theory, there appear two kinds of dualities called **T**- and **S**-dualities[4]. **T**-duality turns a theory with a compactified circular dimension of radius R into one with a compactified dimension of radius $1/R$. **S**-duality goes from a theory with coupling constant c to one with coupling constant $1/c$. By combining **T**- and **S**-dualities one can start from any of the five string theories and reach any of the other four. This led to the conjecture that these string theories are merely different aspects of a single hypothetical theory, the so-called *M-theory*.

There is evidence that M-theory could be an 11-dimensional theory whose effective field theory at large scales is a supersymmetric theory of gravity in 11 dimensions. It has been proposed that the underlying microscopic theory could have membranes as variables, i.e. two-dimensional surfaces. Another proposal consists in reformulating membrane theory as a matrix theory. In spite of interesting leads and indications, however, there is currently no agreement or clear perspective on what M-theory is or if it exists at all.

Another important discovery is the realization that strings necessitate the existence of higher dimensional objects, called *p-branes*. Very roughly speaking, a p-brane can be regarded as an accumulation of large numbers of strings that occupy a part of space. For example, a 2-brane is an object that fills a 2-dimensional surface in space. Mathematically, p-branes have a similar status as the spacetime geometry, since they are exact solutions of the effective field theory. Thus, they are not the result of perturbation theory. Like the geometry and topology of spacetime, p-branes provide a background and strings can be considered as small perturbations or ripples in it. The existence of p-branes also allows for new ways of constructing solutions that are close to the real world. There are species of strings whose ends are restricted to p-branes[5]. From the field theory perspective these correspond to particles that can only move within the p-branes, and they have symmetries of the type encountered in particle physics.

At the time of writing of this article, the most prominent developments in string theory are the AdS/CFT correspondence and the so-called *proliferation of string vacua*. While the discovery of the AdS/CFT duality is important, it appears to be of less direct relevance to the topic discussed here, so I will not say much about it.

---

[4] The letter S happens to be used for both **S**-duality and the action S, but the two concepts are totally unrelated.

[5] *Editor's note*: As in the figure on page 40

The proliferation of string vacua, on the other hand, does have a great bearing on the search for a theory of everything. It raised, in fact, critical questions about whether string theory could at all be a theory of everything. The problem has to do with the way realistic models are constructed from string theory. As explained earlier, the 10 dimensions of string theory are reduced to 4 by making the 6 extra dimensions periodic and small. The specific shape and size of the extra dimensions has influence on what particles the effective 4-dimensional field theory contains and how they interact with each other. Such a spacetime can furthermore be filled with p-branes of various sorts. This affects the particle content and their interactions as well.

It turns out that there are many different ways of compactifying 10 dimensions and many ways to place p-branes in them. So many indeed that one can obtain an enormous variety of different effective field theories, with different natural constants and particles. Each admissible choice can be regarded as a possible vacuum of string theory – hence the term "proliferation of vacua". It is also sometimes referred to as the *string landscape.* It is debated whether all of these vacua are really allowed; since it has not been proven that they are all stable. However, many string theorists seem to take seriously the possibility that there are a huge number of them.

If this is true, compactification and p-branes create the very kind of problem that string theory set out to resolve: the lack of a unique definition of quantum field theory. A string theory in a given vacuum *does* determine the infinitely many coupling constants of its effective quantum field theory uniquely. However, many choices of compactifications and p-branes are available and thus the problem of non-uniqueness reenters through the backdoor. Instead of an infinite number of coupling constants c_i one now has to specify a large amount of information about the shape of the compactified spacetime and the p-branes that reside in it. In a way, the troubles of nonrenormalizability have been shifted from the coupling constants to the properties of spacetime and p -branes.

At this point, I would like to come to the end of this section on string theory and draw a first résumé. What is the message to be taken from this and what is needed in the future?

First of all, I think that string theory has led to very intriguing results. It has pointed out fascinating possibilities of unification and duality, and supported them with concrete calculations that go much further than any other approach to quantum gravity has gone. If it is considered as a candidate for the theory of everything, however, it has serious shortcomings.

What appears most striking to me is that spacetime and strings are treated as separate entities. While string theory manages to unify forces and matter into strings, strings and spacetime remain divorced concepts. To put it briefly: strings move *in* spacetime, but they *are not* spacetime.

This separation is reflected in two ways: it appears firstly in the perturbative formulation. Presently, most string theories are defined as perturbation theories around exact solutions[6]. These exact solutions contain information about the geometry and topology of spacetime. This provides the so-called background spacetime in which strings move. There is clearly a relationship between the two: the motion of strings is restricted and influenced by the spacetime, and the properties of spacetime are to some extent constrained by the presence of the strings. However, the roles spacetime and strings play in this relation are very different. First, one looks at the equations for the spacetime and picks an exact solution. Then, one determines the equations for the strings in this background, and finds a solution for the strings. At no point are spacetime and strings treated as variables on the same footing.

Secondly, even *if* spacetime and strings were treated equally as variables, they would still not be unified in one concept. There is currently no, or no widely favored, idea of what the "mother substance" is from which both spacetime and strings are made. For this reason, string theory is not yet a theory of everything.

This dichotomy between spacetime and strings could be also related to the ambiguities of string theory. As we have seen, quantum field theory suffers from ambiguities because it has an infinite number of coupling constants. Quantum field theory cannot predict the values of these constants, since it is a theory of fields, not a theory of fields *and* coupling constants. The constants are parameters that are chosen by hand.

---

[6] See the remarks on perturbation theory in the previous section.

Something analogous may be happening in string theory. As explained before, the background spacetime is very much like a parameter for the string dynamics. Although it is constrained, there appears to be a large amount of freedom in specifying it. Could it be that the origin of this lies in the conceptual split between spacetime and strings – in the fact that string theory is more of a theory of strings in given backgrounds than a theory of strings *and* backgrounds?

In defense of string theory one should say that there are attempts to go beyond the perturbative scheme: there are examples of transitions between topologies, there is the idea of M-theory as a nonperturbative framework that encompasses perturbative string theories, and the AdS/CFT correspondence which uses a setup where the geometry is only partly fixed by a background. It does not appear, however, that these ideas are currently powerful enough to cure the basic predicament: that string theory does not unify strings and spacetime, and that it does not constrain the spacetime background sufficiently.

The message I take from all this is the following: what string theory calls for is a genuine unification of forces, matter *and* spacetime that is not based on perturbation theory! Moreover, if we want to explain particle physics in terms of extra dimensions and other structures, we need a theory that predicts them stringently! Therefore, a *nonperturbative* theory of fields, spacetime geometry and topology is required. Since it is unified, it would likely be a theory in which there is no spacetime to start with. Spacetime would only arise *a posteriori* through a dynamical mechanism.

LOOP QUANTUM GRAVITY

If we consider nonrenormalizability as the main motivation behind string theory, then it is appropriate to say that perturbation theory and its deficits are the driving force behind *nonperturbative* approaches to quantum gravity.

People who work on them often have a background that is closer to general relativity than to particle physics. The main lesson they learned from general relativity is that spacetime geometry is a dynamical entity. That is, the geometry of spacetime is a variable in the action like any other field, and a large number of different solutions exist. From this perspective, it would seem odd to define a quantum theory of gravity based on perturbation theory: i.e. by choosing an exact solution as a fixed background and to only consider small deviations from it. "Perhaps we can use this as an approximation in certain cases", they would say, "but should we define a fundamental theory like this?" For the same reason, they tend to suspect that the true culprit in perturbative quantum gravity is not so much renormalization, but the perturbation theory by which one computes the renormalized action.

The key question is then: how can one construct a theory of quantum gravity without relying on any perturbative technique?

To describe this issue in more detail, I need to say more about how a quantum theory is defined based on the action function S. The simplest way to explain this is to use the *path integral formulation* of quantum theory. Assume for now that the variable of the theory is the spacetime geometry which is described by a certain field, called the *metric field*. The action is a function of this field.

A central concept in quantum theory is that of the so-called *observables*. These can be any functions of the variables that carry some physical interest: in gravity, this could be the distance between two points, quantities that depend on the curvature, the volume of spacetime *etc.* If it was the volume, for example, the observable would yield the volume of spacetime for any given metric field.

Quantum theory provides a rule for computing the average of such an observable. It is an average over all fields that are weighted in a specific way in the action S. One obtains it by summing over *all* possible values of the field at *all* points of spacetime. The weighting has the effect that some fields are preferred and some are suppressed in the average, and this is how the natural laws are implemented in quantum theory. The preferred fields are the ones that satisfy the natural law according to S, and the suppressed fields are the ones which do not. The hallmark of quantum theory is that this is not a yes or no scheme, but more a matter of degree. Some fields are 100% included and some are 100% excluded, but there are also fields in between that contribute to some extent and have a more ephemeral status in the scheme of things.

The resulting picture is more subtle than the description I gave in section II. There I explained essentially the *classical* approximation of quantum theory, which is good enough in many situations – see (Fig. **3**). For given values of the field at some earlier time, the field at later times is either a solution of the equations or it is not. That is, it is either real or not. This is the classical way of thinking about physics. When quantum theory comes into play, we can no longer say that a particular field is a real one. Many of them contribute and the only thing we can do is to take an observable, i.e. a quantity that we are interested in, and average it over the different possibilities – see (Fig. **3**). As I said before, not all fields contribute in the same way in this average, so there is still a sense of *more real* and *less real* left. However, this sense of reality is more diffuse than in the classical picture.

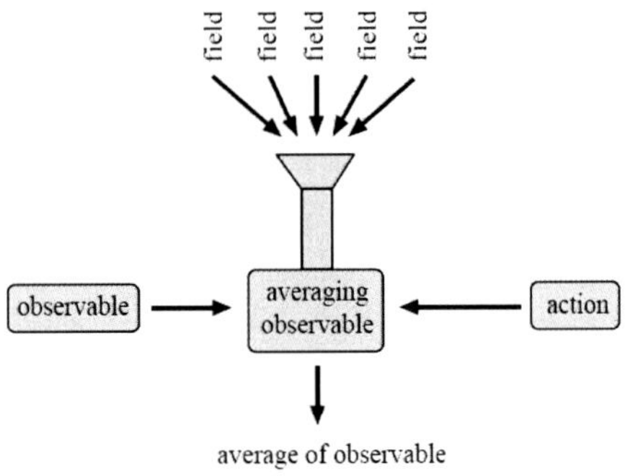

Figure 3: In quantum theory observable quantities are computed as averages over field values. The action enters as a weight in this average.

In short, the meaning of the average is the following: it is the average value of the observable when we average over all fields that are more or less allowed by the laws of nature imposed by the action function *S*. The sum over fields is called a *path integral.* When the first sums of this sort where written down, it was not for a field, but for a particle position that is a function of time. In that case, one had to deal with a sum over all possible histories or paths of the particle, and hence it received the name path integral.

The way I presented it here, all fields appear in the sum and there is no *a priori* distinction between exact solutions (fields that contribute 100%), perturbations around them, and fields which will be greatly suppressed. When the average is meant in this sense, the path integral is nonperturbative. It does not invoke perturbation theory.

The challenge lies in making such an expression mathematically concrete. We know how to define a sum over one variable, or over several variables. But the average over the field is a sum over an *infinite* number of variables – one for each point of spacetime ! How is one supposed to make sense of that? The most simple-minded solution to this problem is to make spacetime discrete; to replace it with a finite number of points, so that only a finite number of variables have to be summed over. There are theories without gravity where this is relatively easy to do. There is a nonperturbative definition of quantum chromodynamics, the theory of the strong force, which uses a spacetime lattice. This lattice consists of small cubes all of which have the same side length.

If gravity is present and lengths and angles are dynamical, discretization has to proceed in more subtle ways. How should one chop up spacetime into pieces when lengths are variable? There are a number of approaches to discretizing gravity that differ in the way they achieve this: *loop quantum gravity*, *causal dynamical triangulations*, *causal sets* and *Regge calculus*.

Here, I will mainly talk about loop quantum gravity. On the side, I will also make a few remarks about causal sets. This does not necessarily mean that loop quantum gravity is the most promising or successful approach to nonperturbative quantum gravity. One reason I chose it for this article are its similarities to gauge theories[7], which makes it easier to put it in a common perspective with string theory and noncommutative geometry. Another reason is simply that I personally worked on it.

[7] The kind of theories that describe the strong and electroweak forces.

This restriction should not convey the idea that other nonperturbative approaches like causal dynamical triangulations or Regge calculus are of lesser importance. Each has its strengths and it is not clear which one will be the most successful in the future.

When it comes to loop quantum gravity, I should emphasize that there are two different ways of defining the theory: there is a formulation where spacetime is sliced into snapshots and temporal evolution is a sequence of such snapshots. This version is called *canonical loop quantum gravity.* The other formulation considers space and time as one block and involves path integrals of the type discussed before. In the latter case, the variables are called *spin foams* – for reasons that will be explained further below. In this article, I will only talk about the path integral or spin foam formulation of loop quantum gravity. In models of spin foams, the discretization of spacetime works as follows: consider, to start with, just a 1-dimensional space, a line, and divide it into intervals. Each interval is characterized by 2 points. Next consider a 2-dimensional plane and divide it up into triangles. Each triangle is determined by 3 points. One dimension higher, in 3 dimensions, we can tile space by tetrahedra, which are fixed by 4 points. When we come to 4 dimensions it becomes harder to visualize things, but the principle is the same. A 4-dimensional space can be subdivided by cells that are specified by 5 points – see (Fig. **4a**). In mathematical terminology, these cells are called *simplices* : the interval is a 1-simplex, the triangle is a 2-simplex, the tetrahedron is a 3-simplex, and in four dimensions they are called 4-simplices. A given way of subdividing a space into simplices is *triangulation*. (Fig. **4b**) shows a possible 2-dimensional triangulation of a rectangle. Since we are dealing with spacetime, we are interested in the triangulation of 4-dimensional spacetime.

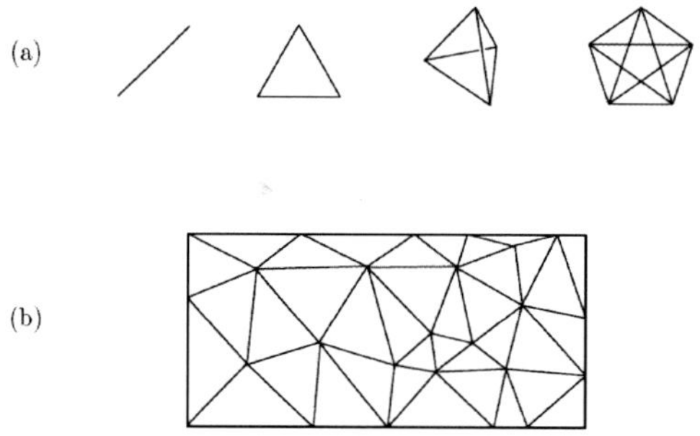

Figure 4: (a) 1-, 2-, 3- and 4-simplex. (b) Triangulation of a rectangle.

I should add at this point that the notion of triangulation does not entail any information about sizes and lengths. We have just subdivided a set of points into subsets of points, but we have not yet assigned lengths, angles and volumes to them. It is the next step to endow the 4-d triangulation with a geometry. To explain how this is done in spin foam models, I have to bring up some more concepts.

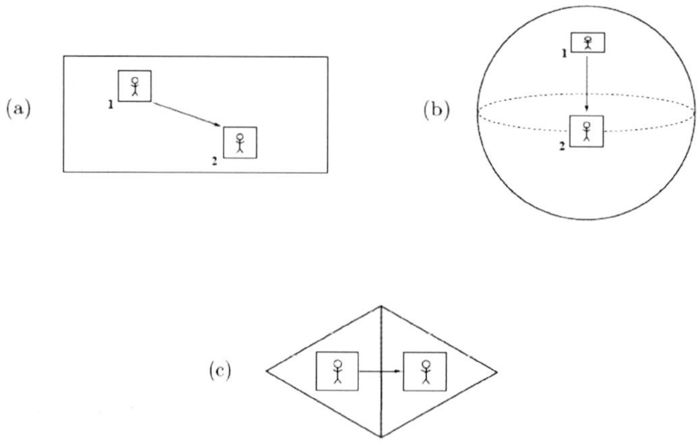

Figure 5: (a) Parallel transport of a figure in a 2-d plane. (b) Parallel transport of a figure on the surface of a sphere. (c) Parallel transport between two adjacent triangles of a 2-dimensional triangulation.

The first one is the notion of an *inertial frame*. An inertial frame can be thought of as the abstraction of a laboratory that moves without acceleration. Mathematically, it is characterized by two pieces of information: a point in spacetime, which represents the location of the lab and a moment in its history; and a spacetime coordinate system around it in which events are registered[8].

Inertial frames are important because in them the laws of nature take their simplest form. Suppose we conducted a little physical experiment in a train that moves with constant velocity. This train approximates an inertial frame and the outcome of the experiment inside it would be independent of the train's speed and where it is heading. Imagine, on the other hand, that we repeated the same experiment in a roller-coaster car. This roller-coaster is not an inertial frame (because it accelerates), and what happens in it would be very much affected by its trajectory. The details of its motion could heavily influence what we measure in the experiment. For this reason, it is much easier to describe the laws of nature in inertial frames, where we can focus on the physical process itself without having to worry about influences from the lab's motion.

There is an even stronger statement behind this: not only are experiments independent of the motion of the inertial frame, they are, in fact, independent of the inertial frame itself. That is, experiments give the same results no matter in which inertial frame they are carried out. This is called the *equivalence principle* and it is a key defining point of general relativity. It means that experiments can never lead to the conclusion that one inertial frame is preferred over another. Inertial frames are equivalent in the sense that the laws of physics are the same in all of them.

Closely associated with this is the notion of a *Lorentz transformation*. Consider two different inertial frames at the same spacetime point: they correspond to two (fictitious) labs that occupy the same position at the same moment, but move relative to each other or are rotated relative to each other. Since the frames occupy the same region at the same time, they register the same physical events. In general, however, the same event will have different coordinates in the two inertial frames. There is a transformation that tells us how to translate the event's coordinates in the first inertial frame into coordinates in the second inertial frame. These transformations are called the *Lorentz transformations*. The laws of physics are the same in different inertial frames, so this means, in particular, that the equations of physics do not change under the Lorentz transformations. This property is referred to as *Lorentz symmetry* or *Lorentz invariance*.

There is also a way to associate the Lorentz transformations to two inertial frames at different spacetime points. Imagine first, for simplicity, a 2-dimensional plane with two inertial frames at points 1 and 2. Connect the two points by a curve or a line. Suppose that we move a little figure at point 1 along the curve to point 2 without changing directions or angles –see (Fig. **5a**). Clearly, lines in the final figure will be parallel to lines in the original. For this reason, the operation of moving a figure in such a way is called a *parallel transport* in geometry.

The concept of parallel transport can also be generalized to more complicated geometries. The next simplest example is the surface of a sphere, as sketched in (Fig. **5b**). As before, one can move a little figure from point 1 to point 2 along a curve. In this case, it is a more subtle question what exactly "parallel" means. I will not go into the details and just say that there is a rule for moving the figure, and it is again called a parallel transport. The important point to note is that in both examples there is a precise prescription for sending points in figure 1 into points in figure 2. If we are given the coordinates of some point in the original figure, parallel transport tells us what the coordinates of the point in the final figure will be.

The same idea also applies when we deal with spacetimes instead of spaces. There, parallel transport specifies a Lorentz transformation that sends coordinates in an inertial frame at point 1 to coordinates in an inertial frame at point 2. Mathematically, the Lorentz transformation is defined by a matrix: the *parallel transport matrix*.

Let us now return to loop quantum gravity and spin foam models. I explained the concept of parallel transports, since these will provide the first set of variables in the spin foam model. Recall that we subdivide spacetime into 4-simplices which are the 4-dimensional analogs of triangles or tetrahedra. To every pair of neighboring 4-simplices we ascribe a parallel transport matrix that translates coordinates in the first 4-simplex to coordinates in the second

---

[8] *Editor's note*: As discussed in "*The Fire in the Equations*".

one. Thus, one can think of each 4-simplex as having its own inertial frame, and the matrix defines a Lorentz transformation from one frame to another. In (Fig. **5c**) this idea is illustrated in a 2-dimensional example.

Another ingredient I need is the notion of *quantum states.* A quantum state is a mathematical object that describes the state of a physical system as a whole, or of a subsystem of it. In the context of spin foam models, quantum states are used to characterize the triangles in the triangulation. Each 4-simplex has 10 triangles on its boundary, and the quantum states specify the area and orientation of these triangles. In the case of spatial triangles[9] the area happens to be determined by a *spin* – a mathematical concept that is also used to describe the internal rotation of particles in quantum mechanics. This explains part of the term *spin foam.* The name *foam* is motivated by the fact that the spins are assigned to triangles, so the whole structure looks a bit like a foam that consists of little surfaces with spins on them.

Together the parallel transport matrices and quantum states form the variables of the model. The path integral is a sum over all possible values of the matrices and states. The observable will be some function of them.

To complete the definition I would also have to explain what the action S is in this case. The action results from multiplying the matrices and states together in a suitable way. The details of this are very technical, so I will not go into them here.

A remarkable aspect of the construction is that one is forced into making spatial areas discrete. The areas are proportional to spins and spins in a quantum theory can only take on half-integer or integer values. This reflects corresponding results in canonical loop quantum gravity which state that area is quantized like spin. This means that an area has to be at least ½ – unless it is zero – so there is a minimum unit of area in the theory.

This suggests that the cells of the spin foam are something like "atoms" of spacetime that cannot be further subdivided. To many such a notion has philosophical and mathematical appeal: continuous space with arbitrarily small distances is mathematically unwieldy, since there is, in principle, no limit to the smallness of structures the theory can harbor. As a result there will be an infinite number of variables. If, on the other hand, there is a minimum unit of length or area, the theory has a built-in limit on resolution, and the number of variables per volume stays finite. In such a scenario, space would be more like a computer screen that has a high, but finite, number of pixels per area. The existence of a minimum area also raises hopes that it could yield a natural solution to the problem of nonrenormalizability. I am skeptical of this claim, however, as I will explain further below.

The strength of loop quantum gravity is in the fact that it attempts to construct a nonperturbative theory from scratch. If this succeeds, one will obtain a single well-defined theory rather than a patchwork of theories that are defined perturbatively. Another nice feature is that it emphasizes the similarity between gravity and gauge theories, i.e. the theories of the electroweak and strong forces. The concept they share is parallel transport. In gravity, it describes the transition between inertial frames and in gauge theories it is the quantity (somewhat more abstract) that underlies the electromagnetic, weak and strong fields. Loop quantum gravity underlines this resemblance by putting the parallel transport centerstage – there are other formulations of quantum gravity that don't.

On the other hand, loop quantum gravity faces substantial criticism that points to existing or potential problems in the theory. One big worry is that the triangulation of spacetime will create preferred inertial frames and break Lorentz symmetry. If we consider a local region of spin foam around some point, it is easy to show that it will look very different depending on the inertial frame we use. A particle at rest with some tetrahedron might therefore interact differently with the surrounding space than a particle that moves at high speed relative to the tetrahedron. Experiments show that deviations from Lorentz symmetry can be at best extremely small, if they exist at all. A theory that tinkers with it treads on dangerous grounds.

This is a generic issue in any theory that uses a triangulation or subdivision of spacetime into cells. There is another approach known as *causal set theory* that tries to avoid these problems by discretizing spacetime in a different way. Here, spacetime is not divided into cells, but instead represented by networks of lines that can be viewed as channels

[9] i.e. triangles whose edges point only in a spatial direction, not a temporal direction.

of cause and effect. It can be shown that such discrete spaces do not break Lorentz symmetry. Their drawback is that it is not yet clear how to define field theories on them.

Another problem in loop quantum gravity is that it is very difficult to determine its behavior at large scales. Perturbative calculations are hard due to the discrete structure of the triangulation. For nonperturbative computations, the natural method would be computer simulations. Unfortunately, these cannot be applied to spin foam models. The reason for this is the so-called *sign problem*. Presently, most path integrals can only be simulated efficiently when one sums over positive numbers. When the numbers are allowed to be negative, the terms in the integral oscillate a lot between positive and negative signs and cancel each other out. As a result, a numerical approximation has to include an enormous number of sample terms in order to approximate the full expression. In some theories, this difficulty can be circumvented by a trick that allows one to replace negative numbers by positive ones. In the case of spin foam models, no such procedure is known.

In the previous section, I tried to explain how string theory succeeds – and yet fails – in resolving the problem of renormalizability. Does loop quantum gravity shed any light on this issue? Some people believe that an upper bound on spatial resolution could cure the difficulties of nonrenormalizability in quantum gravity. I do not share this viewpoint. Problems of renormalization are often confused with problems due to infinities. The two, however, are not the same. It is certainly true that renormalization is easier to understand when a theory has a finite number of variables. But this does not imply that the theory is renormalizable and that the number of terms in the microscopic action S will be small. Currently, there is no reason to expect that the many terms in perturbative quantum gravity will be avoided in loop quantum gravity. Thus, there could appear again a huge number of natural constants whose values are not fixed by the theory.

At the same time, one should keep in mind that the number of complete nonperturbative theories of quantum gravity is exactly zero at the moment. One could say: let us first try to write down *one* and worry about the ambiguities later. I agree with this statement as far as renormalization is concerned.

There is yet another place, however, in which ambiguities appear even before we consider any renormalization. The source of these ambiguities is the triangulation. There are a great number of different triangulations that one could choose from and it is possible that different choices correspond to physically different theories. This could be likened to solids that are made of the same atoms, but differ in their crystalline structure, like graphite and diamond. The triangulation determines which variables belong to neighboring cells, and hence which variables are multiplied with each other in the action. Thus, the ambiguity in the triangulation could also be seen as an ambiguity in the action.

By introducing the discretization of spacetime, one has effectively introduced a whole new layer of structure and information that has to be specified. What are we to do with this? Just pick some triangulation at random and define the theory based on it? As if spacetime was a big crystal whose structure cannot be further explained? Should we choose one from a large number of possible triangulations, similarly as one has to choose from the many possibilities of the string landscape ? Conceptually, this is not a very satisfactory situation. There are essentially two opinions on this issue: one says that we should regard the triangulation of spacetime as a temporary edifice for defining the theory. At the end of the construction one should reduce the side lengths of simplices to zero in an appropriate limit, so that the discreteness of spacetime disappears. This is the viewpoint currently taken in causal dynamical triangulations and in Regge calculus. This approach is completely legitimate, if one's only goal is to define a nonperturbative theory of quantum gravity. It is of less appeal, however, if we aim for unification: it keeps spacetime on one side of the conceptual fence, and matter fields on the other side.

The other possibility is the opposite: namely, that the discretization ought to be viewed as a physical entity in its own right and *not* as a mere auxiliary device. According to this viewpoint, the discrete structure should be part of the dynamics, so that the choice of it results from some physical process; similarly perhaps to crystalline structures emerging from the laws of interactions between atoms. In such a scenario, the discretization would be largely selected by the theory itself, and not by the theorist. Building such a theory is a considerable challenge, but it also offers an opportunity for unification: for putting spacetime and matter on the same footing by considering both as the output of *one* mechanism.

There are budding attempts to extend loop quantum gravity in this direction. In an approach called *group field theory* the triangulation and fields arise from perturbations of a quantum field theory that uses no spacetime to start with. Another project, which is loosely inspired by loop quantum gravity, has been baptized *quantum graphity* : its variables are networks and the aim is to find a dynamical rule (i.e. an action) that favors the formation of graphs which resemble spaces.

At present, group field theory is not very well understood, and quantum graphity is little more than a toy model. I think, however, that attempts along these lines are what is required by the current situation. A discrete spacetime structure has been introduced to make sense of nonperturbative path integrals. The new structure leads to a large amount of ambiguity, suggesting that we need a theory *of* this structure. Since the discretization determines the topology of spacetime[10], this would not only be a theory of quantum geometry, but also of quantum topology. This is, in fact, quite analogous to what I argued for in the case of string theory : in that case it was compactification that resolved a problem and created ambiguities in return. The way out is potentially a theory *of* compactification, and hence of quantum topology.

Does loop quantum gravity also have a chance to unifying gravity with the other forces like string theory ? The short answer to this question is that it is not designed for this at the moment. There is, however, one hint that something like this might be possible in the future. It goes back to the fact that both loop quantum gravity and the other forces are formulated in terms of parallel transports. In the case of gravity, they describe the transition between inertial frames in spacetime. In the case of the other forces, the transformations refer to inertial frames in more abstract internal spaces that do not appear to be extended in the same way spacetime is.

The analogy suggests that both types of spaces may be different realizations of the same concept. Kaluza-Klein compactification is one attempt to formulate this unification, based on continuous spaces. There is another, quite different, strategy to achieve the same goal which involves discrete structures and is therefore more suited in regards to loop quantum gravity. It appears in a theory that is very different at first sight: *noncommutative geometry.*

NONCOMMUTATIVE GEOMETRY

If I had to describe noncommutative geometry in one phrase, I would try: noncommutative geometry defines spacetime by what is *in* it. The statement applies, in fact, to two separate ideas that are used in noncommutative geometry. I will explain them in turn.

The first idea concerns functions which assign numbers to each point in spacetime. Such functions can be added and multiplied with each other. Thus, they form what is called an *algebra.* Each function is an element of the algebra. In physics such functions can represent various physical quantities; in particular, the fields of matter and the fields of the electroweak, strong and gravitational forces. In the following, the word function should be seen as a wildcard for all the possible physical fields and functions that can reside in a given spacetime.

Normally, we would first define the spacetime and then think of functions and fields in it. The idea is to turn this scheme around: to define the algebra first, and to think of spacetime as a second step. The reader may wonder how this is possible: the algebra consists of functions and functions require the spacetime for their definition.

The point is that the definition of an algebra does not require explicit knowledge of what its elements *are*. We only need to know what new element we get when we add or multiply two elements of the algebra. This is sufficient to specify the rules of the algebra.

We could, in fact, give the information about this algebra to another person without telling him what the underlying spacetime is. He would only know that it is an algebra of functions on *some* spacetime. The question is: could he figure out what the spacetime is, based on this "encrypted" information?

It turns out that he could. There is a theorem that states that spacetime can be reconstructed from a suitable algebra of functions on it. More precisely, it is the *topology* of spacetime which can be reconstructed, i.e. the information

[10] i.e. which point is neighboring which other point.

about which point is connected to which other point, the distinction between circle and line, sphere and donut *etc.* So far we are not talking about lengths, volumes and angles. The algebras of conventional spacetimes have the special property that they are *commutative*. This means that it does not matter in which order one multiplies elements of the algebra. An algebra where this does not hold is called *noncommutative*[11].

The above method of identifying spacetimes with algebras brought up the possibility that one could define a more general notion of "space" by using noncommutative algebras. In this case, one cannot interpret the elements of the algebra as functions on a conventional spacetime. However, since we are dealing with an algebra, and commutative algebras correspond to spaces, there may be still enough "spaceness" left in a noncommutative algebra in order to represent some sort of generalized space. Hence one calls the noncommutative algebra a *noncommutative spacetime*. One could say that in a noncommutative spacetime it is no longer possible to separate the space from the fields in it, since the only thing that is left is the algebra. In this way, the concept of spacetime is unified with the concept of field.

The second idea in noncommutative geometry is centered around the notion of resonance. The motto could be "Space is a resonant body" or "Space is defined by its resonances". Consider as a concrete example the drumhead of a drum. After one strikes the drum, the drumhead will resonate and vibrate in patterns called standing waves. What we hear are the vibration frequencies of these standing waves. They are determined by the tension of the drumhead and by its geometry. As in the case of space and algebra, one can turn the tables around and ask if the spectrum of frequencies is enough to determine the properties of the drum. This idea was famously expressed by the question "Can one hear the shape of a drum?" It turns out that one can infer much of the information about the drum from its sound, though not everything.

The branch of mathematics that deals with such questions is called *spectral geometry*. It extends the above idea to space and spacetime in general. The transition from drumhead to spacetime works as follows: the analog of the deformation of the drumhead is a matter field in spacetime. The spectrum of standing waves of the drumhead corresponds to the spectrum of standing waves of the matter field in space. Like the drum, space is a resonant body for the waves and the matter field can only vibrate at certain frequencies determined by the shape of the space. In the presence of gravitational or other force fields the spectrum is modified, since the matter waves oscillate differently under the influence of these forces. In this way, the spectrum also includes information about the parallel transport of gravity and the other forces.

Like the notion of noncommutative space, spectral geometry has a flavor of unification: the definition of space is no longer independent of other concepts. It is characterized by oscillations of matter fields that reside within the space. These oscillations are not solely a property of space (since they require fields), and they are not merely a property of fields (since oscillations require space). The oscillations are the unifying concept that embraces both space and matter fields. Moreover, force fields like the electromagnetic field are represented by specific changes in the spectrum of matter oscillations. Thus, spectral geometry manages to combine the triplet of spacetime, matter and forces into one concept.

Noncommutative geometry is an approach that merges the ideas of noncommutative space and spectral geometry. It uses both an algebra and a spectrum. The advantage of having the two is that they provide enough information to reconstruct the spacetime topology and geometry uniquely. That is, when a suitable commutative algebra and spectrum are given, one can find a unique spacetime that corresponds to this algebra and spectrum. When the algebra is noncommutative, algebra and spectrum represent a more general entity that is called a *noncommutative geometry*. In this case, conventional notions of space or spacetime are no longer valid. Space and fields merge into a greater union that is neither.

The remarkable aspect of this is that it can be applied to the field theories of the Standard Model and gravity. By using the above formalism one can describe spacetime and the electroweak, strong and gravitational fields by *one* algebra and *one* spectrum. Moreover, the action S of the fields can be defined in a very simple manner using the spectrum: it becomes the *spectral action* and equals the number of different frequencies in the spectrum!

---

[11] For example, it does not matter in which order one multiplies ordinary numbers, so $3\times6 = 6\times3$, but rotating an object twice around two different axes would give a different result depending on which rotation is done first! Rotations are noncommutative.

The algebra that is needed for this description is noncommutative. It is only slightly noncommutative, though, or *almost commutative*. As a result, it is possible to give an interpretation of it that is close to a conventional spacetime : it can be thought of as an ordinary spacetime with a layer structure. In 4 dimensions, spacetime extends in the usual way, with its 3 spatial dimensions and 1 temporal dimension. Perpendicular to this each spacetime point splits up into several copies. The steps between these points can be viewed as discrete extra dimensions – as shown in (Fig. **6**). The entire structure looks as if spacetime was thickened, or as if it was made up of layers that are stacked together.

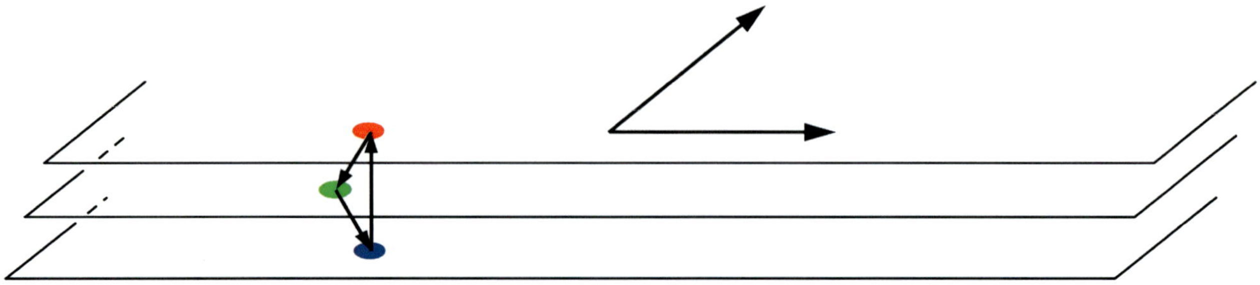

Figure 6: Pictorial representation of an almost commutative spacetime : the long arrows show the extended spacetime directions. Each spacetime point comes in several copies, indicated by colors. The short arrows represent the resulting small extra dimensions.

Due to the presence of the additional dimensions, there are two notions of inertial frame in this spacetime : the usual kind of inertial frame which corresponds to a coordinate system in the "large" space and time dimensions; and another, second inertial frame that describes a coordinate system in the "small" extra dimensions. Accordingly, there exist two types of parallel transports: one that translates between inertial frames of ordinary spacetime, and another one mediating between inertial frames of the small discrete dimensions. The nice thing is that the almost commutative algebra can be chosen suitably, so that the parallel transports of the extra dimensions correspond precisely to those of the electroweak and strong force, while those along the large extended directions become the parallel transports of gravity. Thus, gravity and the other forces appear as different manifestations of the same concept. Their differences result from the fact that spacetime does not have the same size in all directions.

In one way, this kind of unification is similar to Kaluza-Klein compactification in string theory. There are extra dimensions that are small and they are used to explain particular features of particle physics. In another way, however, the scheme of noncommutative geometry is also different and more profound: there is no longer a clear distinction between spacetime and force fields. One could regard the stratification of spacetime as the origin for the electroweak and strong forces. Or one could say that it is these fields that give rise to the layer structure. From the perspective of algebra and spectrum, the distinction between these two statements is a matter of semantics. In this sense, unification á la noncommutative geometry goes further than in string theory, since it unifies spacetime with the forces. There is no force particle or string left that is moving *in* spacetime – the field itself is part of the spacetime structure.

So far I have only stated the positive aspects of noncommutative geometry. Had I explained the mathematics in more detail, it would have become clear that there are also several weak points in the construction. Firstly, it is debatable whether this formulation leads to a true simplification or reduction of information in the Standard Model. While researchers in noncommutative geometry often claim that it does, this is far from obvious from an outsider's perspective. The definition of algebra and spectrum require so many specifications and rules that one may wonder what is really gained by it. Why should one prefer noncommutative geometry over the usual field theory approach when it contains almost as many parameters and arbitrary choices? To put it another way: noncommutative theory does not explain why the algebra and spectrum have the specific form they have: They are neither variables nor solutions to equations. Instead they are posited at the beginning, and then it is shown that they are equivalent to the fields and interactions that we find in the Standard Model and gravity.

My account also glossed over the fact that the unification of gravity and the other forces is not as perfect as one may wish. While all four forces are part of one algebra, gravity enters in a different way than the other fields. The parallel transports of gravity refer to continuous curves in the extended direction of spacetime, whereas those of the other forces arise from paths that can jump between layers. This scheme would be more symmetric if all directions of spacetime were discrete. Then, both the large and small dimensions would be described by the same discrete mathematical structure – they would only differ in how far they extend.

As in the previous sections, I will conclude by asking: what is required to turn this approach into a true theory of everything? Most importantly, one would need a scheme in which the algebra and the spectrum become variables. So far the theory is mainly a dictionary of concepts: it translates the fields and their action into the language of algebra and spectral geometry, but it does not give us any rules for computing and predicting physical processes in this language. For this one has to go back to quantum field theory. So that one can really speak of a physical theory, one would have to have an action which has the algebra and the spectrum as variables, leading to equations that tell us which values of these variables are solutions, and which are not. Hopefully such an action would have the algebra of spacetime, the Standard Model and gravity among its solutions. In a somewhat different context, attempts of this sort have already been made: there are matrix models which have algebras of matrices as solutions, and one tries to recover gauge forces and/or gravity from them[12].

Another issue is, as already mentioned, the integration of gravity with the other forces. Gravity has a special status in noncommutative geometry, since it is associated with the continuous extended dimensions of spacetime, while the forces of the Standard Model arise from small discrete dimensions. To even this inequality out one could try to discretize both the large and small spacetime dimensions. This would render the treatment of gravity and the other forces more symmetric, and make the theory more amenable to a nonperturbative definition.

This thought suggests connections with loop quantum gravity. Noncommutative geometry uses discrete extra dimensions to generate nongravitational forces, while loop quantum gravity employs discrete large dimensions to achieve a nonperturbative definition. By merging these two types of discreteness one could combine the strengths of both theories – if not the theories themselves. In the case of loop quantum gravity, it would offer a strategy for unification: by adding extra layers to the already discrete spacetime, one could recover the other forces in a natural manner.

It is very unclear at the moment how exactly a new loop quantum gravity or a new noncommutative geometry would look like. I will speculate about this in the next section.

FUTURE THEORIES OF EVERYTHING AND THE UNIFICATION OF SPACETIME AND FORCES

In the previous sections, I have reviewed three important approaches in fundamental theoretical physics. The natural question is then: what comes next? We have seen that each of these theories has to struggle with serious difficulties and each of them is far from being a complete theory of everything. Will they transform into new theories, be replaced or merge with each other?

In the following, I want to discuss my personal guess on how future theories of everything will look like. I will speculate on what features such theories could have: what their variables will be, how they will be formulated and what objectives they will achieve. To some extent my guess will be based on my previous review: on what we currently know about string theory, loop quantum gravity, noncommutative geometry and other theories – on the lessons learned from them and their unresolved challenges. On the other hand, it will also be biased by my own preferences and philosophical prejudice and limited by my imagination. As a result, this section will be very speculative and what I will say may well turn out to be wrong. Moreover, it will be very vague and more of a wish list than a concrete suggestion. If I had details, I would already have some proposal for a new theory of everything, but this is not the case.

I will start by formulating four requirements that are motivated by what was discussed in the previous sections. As such these have been stated by various people in different places before, so the elements in this list are not new. However, existing approaches tend to focus on only one or two of these principles at a time, so it may be good to summarize them all in one place.

---

[12] I will say more about matrices in the next section.

1. *Nonperturbative definition:* The basic definition of a future theory of everything will not be based on perturbation theory. It will rather be defined by a path integral that extends over all possible values of the variables and is not restricted to perturbations of exact solutions.

   Comments: Many theoretical physicists would probably agree on this point. As I have explained in section III, string theory starts out with perturbation theory and as a result it is currently more of a patchwork of theories than one single theory. The perturbative formulation also poses a challenge for unification, since it separates the background (spacetime geometry, p-branes etc.) from objects that move in this background. Loop quantum gravity is an approach that attempts to formulate everything nonperturbatively from the beginning.

2. *Discrete structures*: In order to provide a nonperturbative formulation of the theory, one should use finite structures, so that the number of variables will be finite. In this way, one can give a concrete meaning to the notion of a path integral as a sum over all values of the variables.

   Comments: This is an assumption which is certainly debatable. A big concern is whether discrete structures can reproduce Lorentz symmetry. As I have mentioned in section IV, discretizations of spacetime like in loop quantum gravity could disrupt Lorentz symmetry. The results of causal set theory suggest that one can find alternative discrete structures that are manifestly compatible with this symmetry.

3. *Unification of spacetime with forces and matter*: A theory of everything has to unify spacetime with forces and matter. In such a theory, forces and matter are not described by objects *in* spacetime. Instead matter, forces and spacetime derive their existence from a more fundamental entity that encompasses all three.

   Comments: Perhaps this unification will proceed along the lines of noncommutative geometry, where spacetime is defined by the spectrum of matter fields in it and force fields correspond to modifications to this spectrum (see section V).

4. *Space from no space*: The theory should not involve any *a priori* notion of spacetime. Spacetime itself is a variable of the theory, and the equations determine which spacetimes are solutions and which are not.

   Comments: This condition can be seen as a consequence of the previous one. If spacetime is unified with forces and matter, it is a variable like them, and therefore not predetermined. This is also a very radical requirement: it asks us to give up the notion of here and there, of close and far, and assumes that at a fundamental level there is no preferred way in which things are wired with each other. This wiring is only determined *a posteriori* through the interactions of the variables, which is encoded in the equations.

Given these requirements, the first big question is: what are the variables of a theory that can fulfill them? My expectation is that the variables will be *labeled networks* or *large matrices*. Before I give some arguments in favor of this, I will first explain what labeled networks and matrices are, and show that they are closely related.

The abstract notion of a network is very intuitive and easily explained. Consider a set of N nodes and assume that some of them are connected by links as in (Fig. 7). This is a network. When we think of a network in this abstract way, it does not matter where the nodes are positioned. The only thing that is relevant is the information about the connections, i.e. which node is connected with which other node. Thus, we could represent the same network by points in a 2-dimensional, 3-dimensional or d-dimensional space, and all of them would symbolize the same abstract entity. Also, it would not matter if we draw nodes close to each other or far apart: the only notion of distance we have is coming from the links. We can say, for example, that a node neighbors another node if they are connected by a link or that it takes n steps along specific links to go from one node to another.

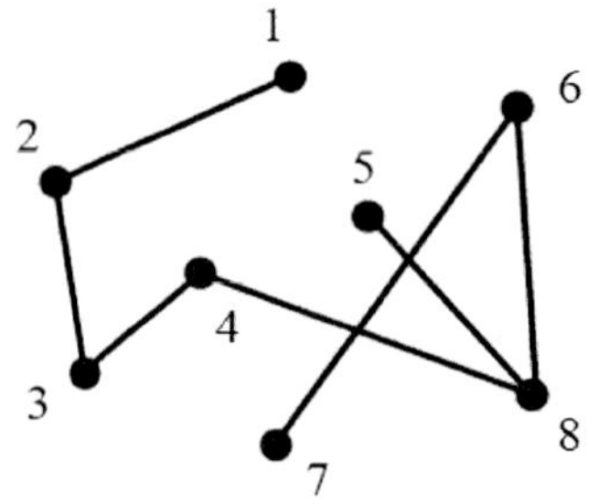

Figure 7: A network whose nodes are labeled by numbers.

A *labeled network* is a network where other mathematical objects are assigned to the nodes and edges/links. The simplest possibility is to attribute numbers to them. We could think of these numbers as variables that live in the "spacetime " defined by the network.

Next I will explain the concept of a matrix. Suppose that we have pairs of numbers i and j that can assume any integer value between 1 and some maximum integer N. If N is equal to 3, we would have the pairs 1 and 1, 1 and 2, 1 and 3, 2 and 2, 2 and 3, 3 and 3, and so on. Given that, a matrix is a rule that assigns a number to each of the possible pairs. For instance, the value 13 assigned to i=1 and j=1, $\sqrt{2}$ to 1 and 2 *etc.* So a matrix is defined by specifying a number for each pair i and j. In order to represent this idea without formulas we can use a diagram: draw N nodes and label them by numbers 1 to N. Connect every node with every other node by a link. These links stand for the different possible pairs i and j. In this diagram, the matrix corresponds to an assignment of numbers to links, as shown in (Fig. **8**). This pictorial way of viewing a matrix also makes clear why matrices and networks are closely related. A matrix can be regarded as a labeled network. The label on the link between node i and j is given by the value of the matrix for the pair i and j. Conversely, a network can be translated into a matrix. The simplest way of doing this works as follows: label the nodes by numbers 1 to N. If there is a link between node i and j, we define the value of the matrix for the pair i and j to be 1, and if there is no link we set it to 0. Thus, the value zero in the matrix corresponds to an absent or switched off link, while the value 1 means that the link is there or switched on.

The next question is then: Why would I think that such networks or matrices could be the right variables for a theory of everything? There are several reasons for this. One is that a network can be viewed as a rudimentary type of space or spacetime. The nodes of the network are the points of the space, and the links specify which points neighbor each other in this space. Since we can use a large number N of nodes and connect them in any way we want, it is possible to generate a great variety of very different spaces from them. Thus, networks or matrices could be suited for constructing a theory in which spacetime is a variable. Moreover, if we restrict ourselves to networks with a finite number of nodes and links, they will naturally lead to a finite number of variables. This is helpful when it comes to defining a nonperturbative theory.

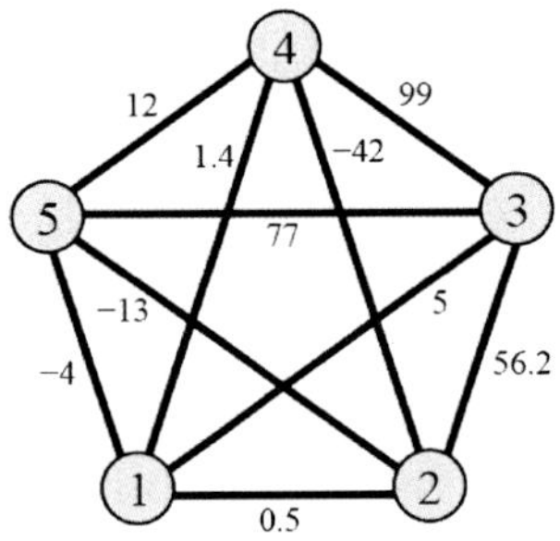

Figure 8: Graphical representation of a matrix: the number on the link between node i and j is the value of the matrix for the pair i and j. To be precise, there should also be links from every node to itself, but these have been omitted for simplicity.

At the same time, networks and matrices bear a close relationship to force fields. Consider a region in a network where we have two sets of n nodes and that the nodes in the first set are linked to nodes in the second set in all possible ways – see (Fig. **9**) Also assume that the links carry numbers as labels. We can interpret this structure as a matrix as follows: label the nodes in the first set by the numbers 1 to n, and do the same for the second set. Then, each link corresponds to a pair of numbers i and j that are integers between 1 and n. Let us define the value of the matrix for the pair i and j to be the number on the link connecting the nodes i and j.

Imagine now that we zoom out a little bit and disregard the details of the little nodes inside the two sets. Then, the two sets look pretty much like two nodes that are connected by a link. In order not to lose the information about the small-scale details we can keep the aforementioned matrix and put it as a label on the link between the two big nodes.

Recall now that force fields are described by parallel transport matrices along curves between points. If we interpret the link between the two nodes as a curve between two points in space, we can view the matrix as the parallel transport between them! Thus, labeled networks naturally give rise to substructures that can be regarded as parallel transports, and hence as force fields. This is a possible route to unifying force fields with the structure of spacetime. It is, in fact, the way unification happens in noncommutative geometry.

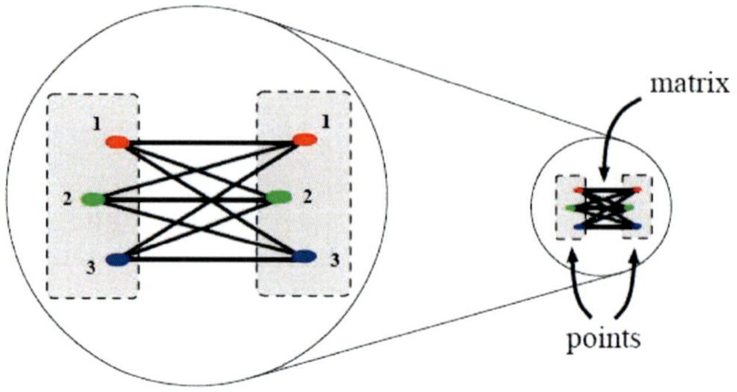

Figure 9: Substructures in labeled networks can be interpreted as parallel transport matrices between points.

A third reason to consider matrices and networks is that they can accommodate the notion of a spectrum. Since a network represents a rudimentary form of spacetime, one can put a matter field in it and analyze its standing waves. Thus, every network, or its associated matrix, determines a spectrum. This opens the way to unification via spectral geometry, where force fields appear as modifications in the spectrum.

We see from this that labeled networks and matrices are quite general and flexible concepts, and that they could provide a conceptual crossing point between spacetime and forces. I expect them to be universal enough to cover a wide array of possibilities. We know, in fact, that networks and matrices underlie a large part of current approaches to fundamental theoretical physics.

The quantum field theories of the electroweak and strong force have parallel transport matrices as their variables. In loop quantum gravity, such matrices are also used as the basic gravitational variables, and they are assigned to links in a discrete spacetime network. Causal set theory involves particular types of networks that can equivalently be described by matrices. The algebras of noncommutative geometry can be defined by matrices[13]. There is even a relation between string theory and matrices, which gave rise to the hope that the hypothetical M-theory is a matrix theory.

The second big question is about the dynamics of future theories of everything. How will their equations be determined? If we believe my guess, the action S will be a function of matrices or labeled networks. But how do we choose this function in order to get spacetimes as solutions? Most networks and matrices do not resemble at all the spacetime of our world. What are the equations that pick out our spacetime from the myriads of possibilities? With this question we enter largely uncharted territory. There are not many hints from the established theories that would tell us how to proceed. In string theory, spacetime is a background and not a variable in the full sense. Loop quantum gravity turns spacetime geometry into a variable, but the discretization of spacetime and its topology are still fixed. Similarly, in noncommutative geometry the algebra and spectrum of spacetime are postulated from the start, and not derived from any dynamical principle.

As previously discussed, there are a few approaches that try to come up with a such a principle. In the context of loop quantum gravity, there is group field theory which attempts to implement a sum over all possible discretizations and spacetime topologies. Quantum graphity seeks a model of networks in which the existence of space depends on temperature: at low temperatures the network crystallizes in the form of a space, similar to freezing water, while at high temperatures this space "dissolves" and the network takes on random forms. Matrix theory, as the name implies, uses matrices and looks for solutions that resemble noncommutative spacetimes.

While all these projects go in the right direction, they have met only limited success so far. What is missing is a paradigmatic example in which spacetime emerges dynamically without us having to put in most of the information

[13] To be precise, some of these matrices are infinitely large, i.e. the number N referred to above is infinite. It may be possible, however, to find approximations in which N is a finite number.

from the beginning. Why is it so difficult to find such a model? The reason for this is probably that we are not used to thinking about physics without spacetime. Most existing theories are about things *in* spacetime, and we have no intuition for physical processes that *generate* spacetime.

I do not know the correct dynamical principle either, but I think that there is a general property that it should satisfy: *the interactions should be local with respect to the network*. So that it becomes clear what I mean by this and why I think it is true, I need to explain the concepts of *locality* and *nonlocality*. We say that something is *local* if it is close to something else that we have in mind. Two points in spacetime are local with respect to each other, when they are close to each other. What precisely is close and what is not can depend on the context, but it always refers to the fact that the distance is smaller than a certain value.

(a)

(b)

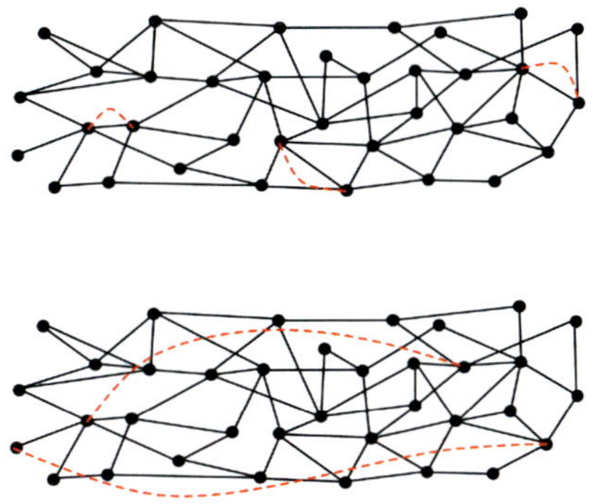

Figure 10: Examples of local interactions (a) and nonlocal interactions (b). Links of the network are shown in black, interactions are indicated by dashed red lines.

The remarkable aspect of the presently known laws of nature is that they are *local* in spacetime. This means that interactions occur only between objects in spacetime that are close to each other. If we were to describe this property in terms of a spacetime network, it would mean that only variables on nearby nodes interact with each other. Influences can propagate over large distances, but this only happens as a result of a chain of interactions between nearby neighbors – see (Fig. **10a**). Imagine, in contrast, a theory where interactions happen between variables that are far apart. Such a theory would be called *nonlocal*, since its interactions do not respect the order of the spacetime network. It would permit immediate interactions between variables that are not directly connected by links of the network – (Fig. **10b**). The locality or nonlocality of a theory is reflected in the form of the action function. The actions of the electroweak, strong and gravitational forces are local in that sense and this is sometimes called the principle of locality in quantum field theory.

When talking about noncommutative geometry I mentioned the idea that spacetime is defined by its resonances. The principle of locality suggests yet another way of characterizing spacetime: spacetime defines which events can interact with each other. Or, conversely, spacetime is *defined* by the interaction of events.

If we take this idea seriously, it can give us a guideline for defining theories where spacetime is a variable and takes on more exotic forms. If the links of the spacetime network specify which sets of variables can interact with each other, this constrains the possible forms of the dynamics. The action function has to be such that interactions are local with respect to the network, i.e. only those variables that are adjacent in the network will interact.

Suppose, on the other hand, that we were to construct a theory where this rule does not hold. In this case, the network would lose much of its meaning, since it no longer tells us which variables directly affect each other. Moreover, we would probably invent another, more convenient network that does reflect the patterns of interactions and allows us to foresee how influences propagate in the system. Thus, we would be led to a theory that is local in this second network, and the original network would have become superfluous. To avoid such a redundancy, it is reasonable to stick to the principle of locality from the start: that is, the network of spacetime should be identical with the network of interactions. This provides us with another take on unification: spacetime *is* the network of interactions.

Since constructing a theory of spacetime is such a hard problem, it would also be a good strategy to start small and simple. That is, before we try to create a theory of everything, we need to look for a simple model that shows that it is indeed possible to start in a formalism without preferred spacetime and to have spacetime arise as a solution to the equations. If a simple model for this is found, it would provide the elementary example of what we want to achieve. I expect this to be the first important step towards a new generation of theories of everything.

Once such an example is available, there would be many interesting things to explore. When we look at the behavior of the theory near an exact solution, it would resemble an ordinary theory in a fixed spacetime : the solution is a particular spacetime and small deviations from it would probably manifest as fields that reside in it. Thus, the solution would determine an effective field theory that describes small perturbations around this spacetime.

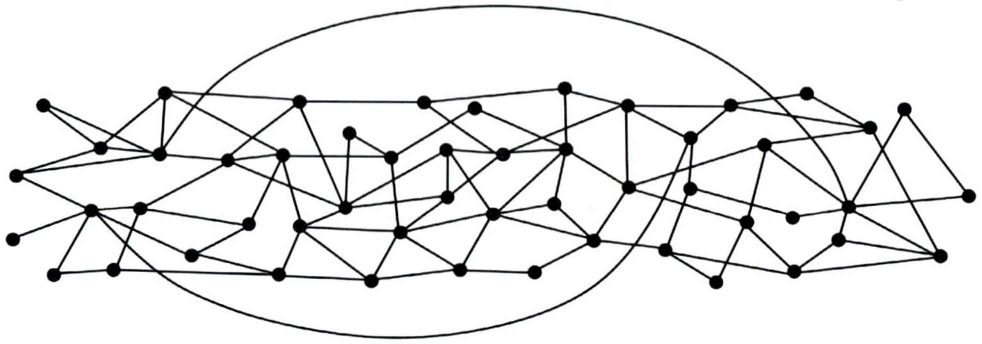

Figure 11: A network that resembles a space. Links between distant nodes constitute defects in the structure of the space.

At a deeper level, on the other hand, the theory would not have any spacetime structure at all: fundamentally there would be no preferred way in which its building blocks are connected with each other. As a result, it is possible that residues of this disorder remain, even when a spacetime structure has formed. In such a situation, most connections in the network or matrix would be such that one can think of it as an ordinary spacetime. Now and then there would be defects, however, that go against the order or topology of this spacetime: i.e. there would be links between points that are too far apart when measured in terms of the other links that comply with the spacetime topology. This idea is illustrated in (Fig. **11**)

How could this type of theory achieve the unification of spacetime, forces and matter? As I explained earlier, substructures of the spacetime network can give rise to matrices that resemble parallel transports of force fields. So that these force fields are everywhere of the same type, the matrices would have to link the same number of nodes at every point – see (Fig. **12**). If this is the case, the spacetime network would have a layered structure as in noncommutative geometry and the forces will result as a consequence.

The question then is whether such a regular microstructure of spacetime arises naturally in solutions of matrix or network theories. One would need, in particular, a solution that resembles the specific spacetime structure of the Standard Model. In order not to run into the same troubles as string theory, the theory would have to predict this structure to a reasonable extent. While many solutions may be possible, the model should not be able to produce anything just because we would like it to. To make an analogy: by combining the atoms of the periodic table of elements one can produce a great number of different chemicals. This number, however, is limited, and the possibilities are not arbitrary. They are constrained in very characteristic ways by atomic physics. Something similar is required of a successful theory of everything: it may allow for a large number of different worlds, but these have to be subject to very specific conditions.

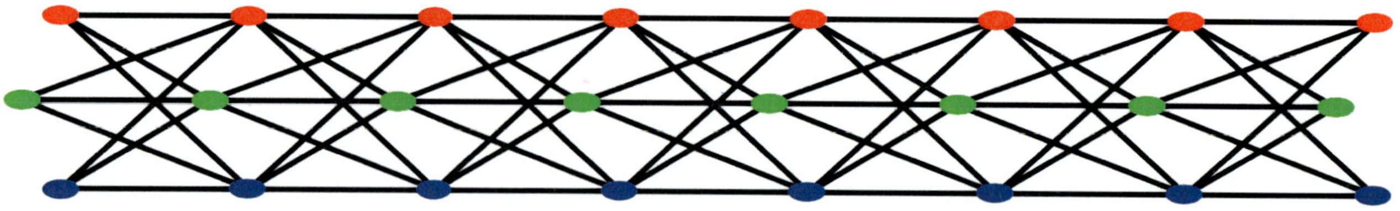

Figure 12: Suitable networks exhibit a layered structure as in almost commutative geometries and could give rise to force fields.

SCIENCE FICTION

In the fifth and fourth centuries BC Leucippus and Democritus promoted the hypothesis that matter consists of indivisible elementary constituents called atoms. Based on this hypothesis they speculated that it is the different packings and scatterings of atoms that are responsible for the existence of different materials and the various sensory perceptions they cause in us. It happens that the basic hypothesis of atomism was correct, and it is also fair to say that some of the speculations derived from it were qualitatively right.

Imagine if the atomists had been even bolder and gone further in their fantasies: perhaps they could have hinted vaguely at the possibility of chemistry, biochemistry, nuclear energy or even nanotechnology. Perhaps it would have led to complete nonsense. Either way they would have pointed out rightly that the idea of atomism has a lot of potential and that many fascinating possibilities are hidden in it.

I want to do something similar here. In sections III to V, I argued that string theory, loop quantum gravity and noncommutative geometry each call for a new theory in which spacetime as a whole is a variable and subject to dynamical equations. If we believe in unification, and that spacetime has to be unified with matter and forces, this hypothesis is reasonable. In the previous section, I explained this hypothesis in more detail, and vaguely outlined a framework in which it could be realized mathematically.

In this section, I want to speculate further on the possibilities that are hidden in this basic premise. Although the correctness of such speculation is highly uncertain, it can nevertheless serve a purpose: it will remind us that the space of possibilities is big. The human mind tends to assume that what it knows is almost all there is, or at least, that the unknown is an extrapolation of what is already known. The historical truth is, however, that the unknown part of the world has always been spectacularly bigger and beyond what was known and understood. I assume that this is still true today. For this reason, wild speculation is good. It allows us to conceive of the possibility that the world is very different from what we think it is.

Let me therefore ask the question: what if the hypothesis is true and we live in a world where spacetime is not absolute, but is rather the product of a dynamical process? My speculation on this consists of two parts: first I will consider the possibility that a theory of everything gives rise to not only one space, but several spaces at once. Secondly, I will argue that a theory of variable spacetime is close to being a theory in which the laws of nature themselves are dynamical.

If we buy into the idea that there is a theory of everything that generates spacetime, it is conceivable that it could also produce several spaces or several topologies as solutions. Consider for simplicity the case where there appear, say, two spaces: space 1 and space 2. In the picture of networks that I was entertaining in the previous section this could mean, for example, that the nodes split into two sets – see (Fig. **13a**) – and most links connect only the nodes of space 1 or only the nodes of space 2 respectively. Another, more subtle possibility would be that each of the two spaces occupies the same set of nodes, but the links fall into two categories: one set of links characterizes space 1 and another set space 2 – (Fig. **13b**). The third possibility is a combination of the first two, where there is only partial overlap between the nodes of the two spaces – (Fig. **13c**).

(a)

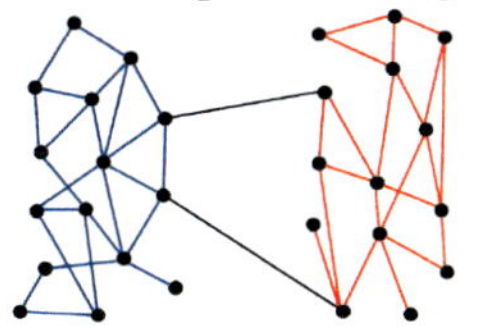

(b)

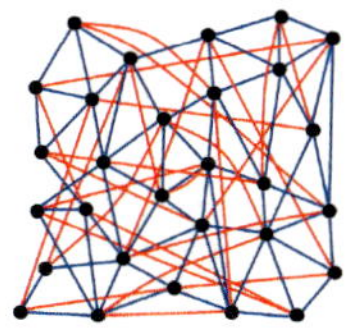

(c)

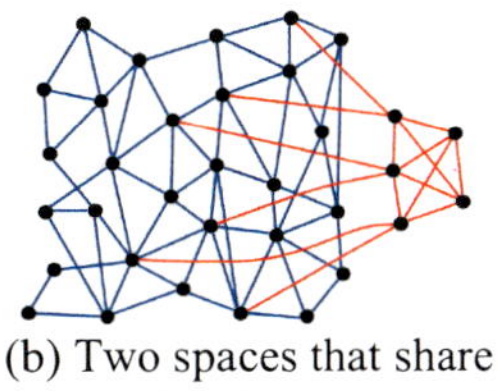

Figure 13: (a) Two spaces that are almost disconnected. (b) Two spaces that share the same set of nodes. The blue links represent a 2-dimensional space, while the red links form a circular chain. (c) Two spaces whose nodes partially overlap.

Imagine now that the links of these spaces also carry labels, so that we can accommodate parallel transports of force fields. Matter fields could be represented by labels on the nodes. In the first scenario each of the two spaces would be governed, to a first approximation, by an effective field theory that can be viewed as a theory on one space alone. However, since there are nonlocal defects that go between the two spaces (as discussed in relation to (Fig. **14**), there would be deviations from this description. There would occur some interaction or information exchange between the two effective theories on spaces 1 and 2.

In the second scenario, both spaces would operate on the same set of nodes. According to what I said earlier, links determine which variables interact with which other variables. Since we have two different types of grid on the same set of nodes, one could think of this as two different kinds of interactions that compete with one another. Depending on the specific values of the coupling constants it may happen that one spacetime dominates over the other, so that the interactions along the second spacetime have hardly any influence. In the third scenario, space 2 is a network that is partially independent and partly connected to the network of the bigger space 1.

At this point, you may object and ask whether these ideas are at all consistent with what we observe. If these spaces existed, we should have noticed them already! This is, of course, a possibility, and in that case my speculations would simply be wrong. It is also possible, however, that these other networks exist, but that they have gone unnoticed because we have not yet asked the right questions. Current scientific thinking is usually based on the idea that there are laws of nature at small scales of spacetime and that these completely determine the dynamics of the world at every level. A theory in which spacetime is emergent naturally suggests that there could also be alternatives to this scheme. As explained before, one could imagine systems in which several networks compete with each other and in such a situation it is no longer true that the entire system can be reduced to the microscopic structure of one network. It may be that one of the existing networks is dominant, but there would nevertheless be small deviations from it, due to the presence of other networks. Thus, while reductionism is a good approximation, further, more refined investigations might show that the world is actually not reducible to one network, and that it has more of a multipolar structure.

The idea of competing networks brings me to the second speculative idea. I argued before that spacetime determines which variables interact with which other variables. So, if spacetime is a variable, it means, in a way, that the interactions are variables, too. In other words, we are only a short step away from turning the laws of nature themselves into dynamic entities. So that they become completely dynamical, it would suffice to convert the natural constants, like charges and masses, into variables as well.

Note, however, that whenever something becomes relative, there remains something else which is absolute. In the case of special relativity, time intervals and spatial distances became relative, but they were replaced by an absolute concept: the geometry of spacetime. Similarly, making the laws of nature themselves dynamic would not imply that the resulting theory is void of any structure. There would have to be a meta law which governs how the spacetime network and the associated natural constants evolve.

There have been previous speculations about the evolution of the laws of nature. One proposal says that the formation of every black hole causes the creation of a new baby universe with modified laws of nature. The iteration of this process will result in a Darwinian evolution of universes, since some universes are more suitable for the production of black holes than others. Another scenario, going under the name of *eternal inflation*, says that now and then a new universe with a new set of natural constants inflates within a given universe. Since this mechanism repeats forever, we end up with an infinite tree of worlds along which the laws of nature evolve.

What I want to emphasize here is that a theory of spacetime topology is *already* a theory of evolving natural laws. If spacetime is dynamic and equivalent to the network of interactions, the interactions are dynamic, too. And these interactions constitute the laws of nature, since they correspond to the terms in the action function.

Ordinarily we think of the state of a system as something relative: the particular way in which the atoms are arranged in a room or in the solar system has many arbitrary and complicated features that are not predicted by the fundamental laws of physics. The philosophical term for this kind of randomness is *contingency*. The laws of nature, on the other hand, are considered as absolute and immutable.

A scheme in which the laws of nature change suggests that the present state of these laws could have contingent features as well. The state of matter and forces that we observe in our immediate environment is very complex, and it is the product of a very long and involved evolution. Could it be that the laws of nature, too, are complex and the product of a long process? If this was true, the currently known laws of physics may turn out to be just the simplest layer of a much richer structure. That is, the known terms in the action of field theories would have to be supplemented by many additional terms that correspond to more complicated types of interactions. Moreover, these forms of interaction might not be static, but still undergoing change.

The combination of temporal evolution and contingency led to an extraordinary degree of complexity and organization in matter and force fields. The pinnacle of this evolution is the biological world. Given this, one may wonder if a similar evolution could have also led to higher levels of organization in the structure of spacetime and the physical interactions. Could reality, at its very basis, exhibit emergent features?

NONPERTURBATIVE RENORMALIZATION

In this last section, we will go full circle and come back to the topic with which the article began. I started out by explaining nonrenormalizability in quantum field theory and why it is a problem: that there is an infinite number of natural constants and that they are not predicted by the theory. I then went on a long discourse about theories of everything, which led me eventually to the speculation that the laws of nature could be contingent. These two statements are related. Both suggest, based on very different arguments, that the structure of the laws of nature is complex and that it has many degrees of freedom.

The nonrenormalizability of perturbative quantum gravity says that it is not consistent to assume that there is only a finite number of terms in the action. This would be like attempting to find the square root of 2 by using only numbers with a finite number of digits. If one tries to do so, one would always find some discrepancy between the square of the proposed number and 2. However, with an infinite number of digits, it works. In the context of renormalization, the counterpart of the condition $x^2 = 2$, would be some deeper theory that determines the infinite number of constants in quantum gravity.

The argument for the contingency of the laws of nature was more abstract. It said that spacetime should be equivalent to the network of interactions, and, if spacetime is variable, the allowed forms of interactions will be dynamic as well. Thus, a long series of terms in the action could arise as a consequence of contingency. In terms of the analogy with numbers, this would be similar to the number that describes the magnetic field at some location on the earth's surface today. It contains a large, if not infinite, number of digits. To some degree, it can be explained by the laws of physics, but it also exhibits a lot of accidental features that could never be predicted by physics.

There is another approach in theoretical physics that takes the possibility of infinitely many natural constants seriously and which I have not mentioned so far. It is the theory of *nonperturbative renormalization.* In the context of quantum gravity it is also referred to as the *asymptotic safety scenario*. This idea offers another solution to the issue of the indeterminacy of infinitely many constants. The claim is that there may be a natural way for determining these constants when renormalization is done *without* perturbation theory. It is therefore a nonperturbative approach like loop quantum gravity. Nonperturbative calculations suggest that the infinitely many constants are actually strongly constrained and to a large extent fixed. The name for this scenario is asymptotic safety, since in the limit of infinitely small scales the values of the natural constants settle into a specific value. Thus, one is "safe" from the troubles of perturbative nonrenormalizability. When this happens, one speaks of *nonperturbative renormalizability.*

In practice, computations can only be done with a finite number of terms and this has been subject to criticism. It is a bit as if the nonperturbative renormalization gave a way to compute the first n digits of $\sqrt{2}$. First one does a calculation with 2 digits, giving 1.4, then, one does it for 4 digits resulting in 1.414 and so on. Each calculation involves only a finite number of digits following the decimal point. The critics argue that this does not guarantee that there actually exists a number $\sqrt{2}$, and that numbers with infinitely many decimals could be a totally different story. The people behind nonperturbative renormalization point out, on the other hand, that computations with more digits confirm the first few digits that were found in earlier calculations. This supports that $\sqrt{2}$ really exists, and that one gets closer and closer to its value as the number of digits is increased.

Let us push this analogy even further and consider the line of real numbers. Each such number (such as 49.5 or $\sqrt{2}$) corresponds to a finite or infinite sequence of natural constants which determines an action and associated laws of physics. Thus, the real line (the set of all real numbers) would be the same as the set of all possible laws of nature. We can think of this set as a space, and each point in it specifies the coupling constants c_1, c_2, c_3 *etc.* and hence the allowed interactions. This is the space that nonperturbative renormalization deals with and the aim is to compute our location in it.

In the light of the previous sections, this gives us another hint at what a theory of everything will be. If we identify spacetime with the network of interactions, each choice of these interactions corresponds to a possible spacetime. From this perspective, non-perturbative renormalization can be viewed as an attempt to determine this spacetime-interaction network from first principles! It does so, however, in a very restricted sense, since the set of interaction terms is very limited in practice.

For a theory to be a true theory of everything one would have to make the concept of spacetime and the laws of nature sufficiently general. That is, the notion of spacetime has to be flexible enough to accommodate the many ways in which spacetime can be wired. Equivalently, the space of coupling constants has to be large enough to account for the many ways in which variables can interact with each other. My guess is that labeled networks or matrices provide a suitable language for this. In this picture, loop quantum gravity could be regarded as a first guess at what spacetime and interactions are when represented as a network; as if one estimated $\sqrt{2}$ to be 1.41. True spacetime is likely to be more complicated and subtle, like the numbers $\sqrt{2}$, e or π, but it is a first step.

REFERENCES

[1] B. Zwiebach, *A First Course in String Theory*. Cambridge University Press, 2009.
[2] C. Rovelli, *Quantum Gravity*. Cambridge University Press, 2007.
[3] T. Schucker, *Forces from Connes' geometry,* Lec. Notes in Physics, p659, 2005.
[4] M. Niedermaier and M. Reuter, "The asymptotic safety scenario in quantum gravity", *Liv. Rev. in Relativity,* 9, 2006.

CHAPTER 6

A Vision of Quantum Gravity

Tim A. Koslowski

Perimeter Institute for Theoretical Physics, Waterloo, Ontario, Canada

Abstract: In this article, I discuss my view of the problem of quantum gravity. I then explain two approaches to solving this problem: Loop quantum gravity and the asymptotic safety scenario, which tackle the problem from opposite ends. Loop quantum gravity tries to describe the fundamental building blocks of spacetime with the hope that the resulting theory may explain gravity as we know it. On the other hand the asymptotic safety scenario considers theories that are as close to gravity as a quantum theory could be and asks whether this setup satisfies the conditions of a fundamental theory. While loop quantum gravity is in a sense a bottom-up approach, the asymptotic safety scenario is a top-down approach to quantum gravity. I explain why the weaknesses of either approach are the strengths of the other, suggesting the main idea of this article; that a merging of both approaches should yield fundamental insight into the problem of quantum gravity. I discuss what the consequences of such a merging could be.

I. INTRODUCTION

One of the major open problems in theoretical physics is to find a theory of quantum gravity. It turns out that the main obstacle to this is to achieve predictivity. This means that we would like to extract the undetermined part of the theory from a small number of experiments, and after these experiments are done, we should, in principle, be able to use the theory to predict the results of any new experiment. In contrast, if we accepted the most general mathematically consistent theory, then whenever we started a new kind of experiment, we enter into a domain of the theory that is not yet determined and we cannot predict what is going to happen. So, predictivity in this context means that the theory is more than a sophisticated devise to store experimental results.

The failure of many well-understood concepts in physics to yield a satisfying solution to this problem has fueled research in diverse directions, most prominently string theory and on fundamental questions of quantum theory. These two directions are liberal in the sense that they introduce new concepts into physics. In addition, there are more conservative approaches to quantum gravity that "only use concepts that were in some sense already there". The two most developed conservative approaches to quantum gravity are the bottom-up framework of *loop quantum gravity* and the top-down approach of the *asymptotic safety scenario*.

Loop quantum gravity is based on a very strict interpretation of a mathematical framework known as canonical quantization and is concerned with constructing a mathematically precise description of the fundamental building blocks of spacetime. Canonical quantization can be formulated as a mathematical program and it has been at the core of loop quantum gravity to strictly follow it and formulate the central results as mathematically proven theorems. Although these statements are proven in a particular mathematical setting, they do not prove that this setting is actually realized in nature. Moreover, there is currently no experiment with the power to test the setting itself, meaning that loop quantum gravity, much like string theory and other competitors, is a speculation. There is, however, an even more serious problem: that it is not clear that the collective behavior of the fundamental building blocks of nature that loop quantum gravity assumes do explain large scale spacetime as we know it. Let me explain this with an example: The simplest atom is the hydrogen atom and both theoretical physicists and theoretical chemists have learned a lot about atoms and chemistry from studying the hydrogen atom, but despite all the lessons that can be learned from this, they are insufficient to explain most chemical reactions, because these involve different atoms. Therefore to describe real chemistry, we have to embed our knowledge about the hydrogen atom into a larger theory that describes all stable atoms in the periodic table. This leads to one way of stating the main idea of this article: Loop quantum gravity may teach us important lessons about the *nature* of quantum gravity, but in order to make it work as a *theory* of quantum gravity, we have to embed it into a framework that describes large scale spacetime as we know it.

The asymptotic safety scenario, on the other hand, is based on a very different premise: The observation that it is very often sufficient to look at large scale physical systems neglecting the effects of all small scale details of the system. The effective theory for the system seen at a coarse zoom level is related to the fine theory that contains all the small scale details by an equation called the *renormalization group equation*. As I will explain later, it turns out that it is possible to find existence conditions for a predictive theory by just considering the renormalization group equation itself. This is called "asymptotic safety" and is hoped to be realized in a theory of quantum gravity. It is, however, very difficult to practically check the asymptotic safety conditions, so in order to find them it is very useful to have a guess at what the predictive theory could look like. This leads to the second way of stating the main idea of this article: We may use loop quantum gravity to guess at the asymptotic safety conditions.

This article is organized as follows: In the next section I will explain what the problem of quantum gravity is in my view; to do this I introduce the main concepts of this field of physics. Thereafter I will introduce the framework of loop quantum gravity and the asymptotic safety scenario, by explaining the respective setup, the main achievements and the open problems of both approaches. I conclude by arguing that a merging of both approaches would yield new fundamental insights and speculate on what these might be. I will end each section with a short summary of its essential points.

Before I start I have to add two disclaimers: First, I have to warn the reader that the entire field of quantum gravity is speculative, because there is currently no direct experimental evidence in support of any theory of quantum gravity. The motivation for the search for such a theory is that experiments tell us on the one hand that the world is fundamentally described by a quantum theory and on the other that gravity is a force of nature. Second, since this article is intended for a general audience, I am not going to provide references to original literature. This would be virtually impossible, since many single sentence statements in this chapter emerged as the essence of hundreds of original papers. I will also not use particular researchers' names unless it has become the customary name of an idea or concept. It should however be clear that the theories that I am discussing are not a consequence of my own work, but rather of a community of hundreds of researchers.

WHAT IS THE PROBLEM OF QUANTUM GRAVITY?

To be able to explain the problem of quantum gravity, I have to take the reader on a short and rather cartoon-style tour through theoretical physics. I will only introduce concepts that are important for the subsequent discussions. The names of the concepts are in italics.

Physicists like to discuss physical systems in terms of *actions*. To understand the concept of an action, consider soap bubbles: We know from observation that freely floating soap bubbles are spherical in shape and the only thing that distinguishes them is the volume of the air inside the bubble. Spheres on the other hand can be characterized as the surfaces with the smallest surface area enclosing a given volume. We can thus express the physical law of bubble shapes as follows: "The shape of a freely floating soap bubble that encloses a given volume of air is such that the surface area of the soap bubble is minimal". At this point it is awkward to replace the simple intuitive concept of a sphere with the more complicated abstract concept of a minimal surface area. To explain why it is nevertheless a good idea to do so, we consider a soap bubble that is squeezed into some wire structure frame. Its shape is no longer a sphere, but rather a very complicated surface whose details depend on the details of the frame itself. There is no way of expressing the law of shapes in the form: "Given frame A the shape of the bubble is B". However, it turns out that the law that determines the shape of the soap bubble can still be stated as: "The shapes of squeezed soap bubbles that enclose a given volume of air are such that the surface areas of the soap bubbles are minimal". Although more abstract, this is a very efficient way of writing the law of shapes. It turns out that similar efficient formulations of the laws of physics can be found for most physical systems. These laws are of the form: "Given some condition (in our example the enclosed volume and the shape of the frame), the state of the system (in our example the shape of the bubble) is such that some quantity (in our example the surface area) is minimized". The quantity whose minimization yields the dynamical laws is the so-called action, moreover it turns out that the description of physical systems in terms of actions is in almost all cases very efficient, which is why physicists like to use them.

Having understood the importance of actions for theoretical physics, one is lead to wonder how physicists determine the action of a system. In general one is unable to give an algorithm for the identification of an action for a physical system, but there are two guiding principles: First the determination of the observables of the system and second the determination of the *symmetries* of the system. Symmetries turn out to be powerful tools: Consider for example that somebody told us that he drew a closed shape on a sheet of paper, then there is an infinity of possibilities e.g. a triangle, a square, a rectangle, a circle or an ellipsoid. However, if the artist told us that the shape looks exactly the same after an arbitrary number of rotations of the paper, then we can unambiguously deduce that the shape is a circle. A "symmetry" is a statement of the type: "An observable quantity (in this example the shape) does not change under some transformation (in this example the rotations)." This is an example of the power of symmetries.

Most physicists believe in relativistic *spacetime.* It is a counterintuitive concept in the sense that in everyday life we experience distances in space and distances in time (i.e. durations or periods) as two very different quantities. It does however turn out that space and time are two aspects of the same thing: spacetime, i.e. a bundle that consists of both space and time; its constituents are hence neither points in space nor instants in time but *events* that take place at one point in space and time. The difference between space and time is then encoded in the interval between events, which may be decomposed into two pieces: a spatial distance measured as a positive value and a time period as negative. So far it may seem to be an awkward bundling of space and time. But the concept goes deeper: It turns out that this awkward interval that can be positive as well as negative is the quantity that provides the *action* for the motion of free particles. Hence the law can be stated as: "Free particles move from event A to event B along minimum length curves in spacetime". But the bundling of space and time to spacetime goes even further: It turns out that the symmetries of the spacetime interval are more than just the symmetries of spatial distances plus the symmetries of time periods, but that there are new transformations that transform a bit of space into time and a bit of time into space, such that the net effect leaves the spacetime interval unaffected. These so-called *Lorentz transformations*, mixing space and time, are at the heart of special relativity.

Next, consider the so-called *least action principle*: "A particle moves from point A to point B along the shortest curve connecting A and B". Let us suppose that this applies to particles on a flat surface, e.g. a desk, then all particles move along straight lines, because these are the shortest connections between the two points A and B. However, if we apply this principle to the motion on the surface of, say, a coffee mug then we observe that a particle moving from one point on the rim to the opposite point on the rim does so not along a straight line, but along the rim itself, traversing a circle, because this is now the shortest curve connecting the two points (the shortest *on* the surface). This is of course due to the fact that the surface of the mug is not flat. Now suppose that the mug is rendered invisible, and we can only see the particle moving on a circular trajectory, then we may think that the motion is due to the action of a force instead of attributing it to the curvature of the mug's surface. It does indeed turn out that gravity can be described as a force and equivalently as curvature, however not as a curvature of space as in the mug example but as a curvature of spacetime. This description of gravity is a central building block of general relativity. Gravity, in the language of general relativity, is nothing more but the deviation of the geometry of spacetime from flat geometry – see (Fig. **1**).

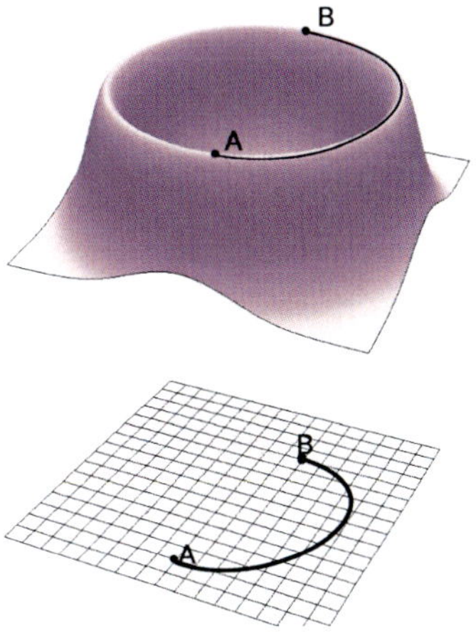

Figure 1: Top: A surface similar to that of a coffee mug. A particle moves along the shortest curve from A to B, near the rim. Bottom: The same curve seen without the curvature of the mug's surface. The particle moves along a bent curve, seemingly under the influence of an unseen force.

Moreover it turns out that the action principle that describes gravity has a very simple form in the language of geometry: "Spacetime geometry is such that a quantity, called the *total scalar curvature*, is extremal (i.e. either minimal or maximal)". One can think of this scalar curvature as a particular mathematical generalization of the inverse of the radius of a sphere: The neighborhood of a point on a sphere with a small radius appears much more curved than the neighborhood of a point on a sphere with a larger radius. The total scalar curvature is then obtained by summing the scalar curvatures of all neighborhoods. One particular special case of this is flat spacetime, which exhibits no gravity; another famous solution is a black hole, which exhibits an attractive gravitational force towards its center.

The geometric language turns out to be very efficient in describing gravity, but it comes at a price: The mathematical description of geometry is very redundant. For example, the concept of individual points does not have a meaning in geometry, only the relationship between points do, such as the distance between two points. In practice one assigns labels (coordinates) to points and expresses geometry in terms of relationships between these labels, which are interpreted as relationships between the points that the labels are assigned to. But any relabeling of these points still describes the same geometry; this is the redundancy in the description of geometry. There is, however, a way of obtaining true geometry from this redundant description. It follows from the following idea: Geometry is independent of relabeling, so if we device a hypothetical mathematical transformation that performs a relabeling then it does not alter the geometry. But this is nothing else but to say that the relabeling transformations are symmetries of the geometry and therefore of gravity. This means that we can specify a geometry by presenting it in one of the infinitely many redundant ways, if we keep in mind that the geometric quantities that we calculate from it are invariant under the relabeling symmetry. The technical term for the relabeling is *diffeomorphism* and it is a special case of the more general concept of *gauge symmetry*.

Let us now introduce the "*quantum*" part of quantum gravity: It is rather easy to explain quantum theories in terms of the mathematical structure that is used to define them, but the interpretation of these mathematical structures is still ongoing. There are therefore many different ways to paint a cartoon description of quantum theories and each has its own advantages; the one I use here is adapted for the purpose of explaining the specific problem at hand; that of quantum gravity.

Consider a classical point-like particle: we can describe it by its position and velocity (or more conveniently its *momentum*, which is, classically, the product of its mass and its velocity). This data, the position together with momentum, is called the *state* of the particle. In general, a state is a set of data that completely describes the system, meaning that every other quantity pertaining to the system can be calculated from this data. The fundamental observation of quantum mechanics is that in the microscopic world one has to replace the state with a mathematical construct known as the *wave function*. This means that in order to understand the behavior of a microscopic particle one has to describe it as a wave, like the one that propagates on the surface of an otherwise silent pond after we toss a stone into the water. The wave carries energy, so for example if we have a ball floating on the pond then this ball will start to move when the wave reaches it. It turns out that the mean energy carried by a wave is proportional to the square of the amplitude of the wave, i.e. if the peaks of one wave are twice as high as the peaks of another wave then the energy density carried by the first wave is four times the energy density carried by the second. It also turns out that the *probability* of finding a microscopic particle at a certain point is analogous to the energy density carried by the water wave in that it is proportional to the square of the (modulus of the) amplitude of the wave function.

A wave is however not completely determined by its amplitude, but also by its so-called *phase*. To understand this it is instructive to consider a single point that is affected by the wave. This point will undergo periodic motion similar to that of a pendulum. If we compare two pendulums then we can compare the peak heights (amplitudes) but even if two pendulums have the same amplitude they may not be in the same state because they may be at different positions in their periodic motion. This difference in position is called the phase of the wave. Many of the intriguing features of quantum mechanics are due to the wave function's phase. If we reconsider the silent pond and toss two stones then two waves will propagate on the surface of the pond and when the two waves overlap they will interfere (the technical term is interference as well) with one another: at certain points the two waves will enhance each other while in other places they will exactly cancel one another. Whether constructive enhancement or destructive cancelation occurs is due to the phase. If both waves have the same phase then their amplitudes will add up, because a peak of one wave will always meet a peak of the other; if both waves have a maximal difference in phase then a peak of one will meet a valley of the other and vice versa, always leading to cancelation. It turns out that the wave functions that describe quantum mechanical particles exhibit analogous behavior – see (Fig. **2**).

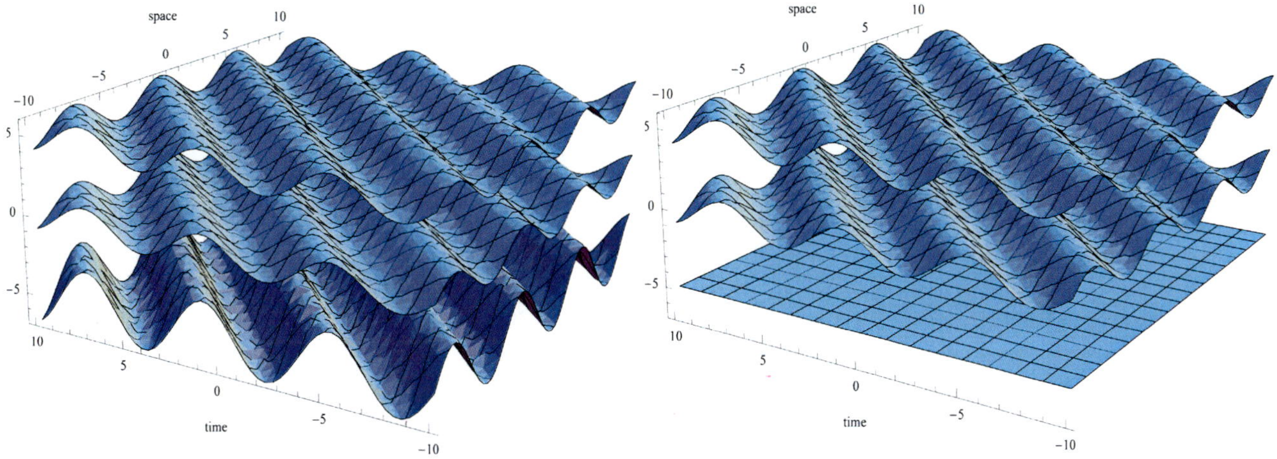

Figure 2: Left: A peak of the top wave is always met by a peak of the middle wave and a valley is always met by a valley. The bottom shows the sum of both, enhancing each other (constructive interference). Right: A peak of the top wave is always met by a valley of the middle wave and vice versa. The bottom shows the sum of both waves, canceling each other out (destructive interference).

In the case of a classical particle it was clear that the relevant information is its position and momentum. But what is the information about a quantum particle whose state is described by a wave function? I mentioned that the probability of finding a particle at a given position is calculated from the wave function, so in general we cannot answer the question: "Where is the particle?" Instead, we have to rephrase the question into "What is the *probability* of finding the particle in a certain region?" The same is true for the particle's momentum. The general question that one can ask about a quantum system has the form: "What is the probability that a measurement (such as the position of a point particle) yields a certain result (such as the particle being in a certain region)?" Moreover, it turns out that the description of a system in terms of correlations of measurements is very well-adapted to quantum systems. Correlations answer questions of the kind: "Given a measurement A (e.g. a particle is at an initial time in region A), what is the probability for another measurement (e.g. the particle is in region B at a given later time)?" This type of question is typical for another type of systems known as statistical systems. It turns out that in many cases there is a precise mathematical way to map correlations of quantum systems into correlations of statistical systems. This mathematical link between certain statistical and quantum systems provides a source of significant insight into the nature of quantum systems.

Let me explain what a statistical system is using an example: Consider an ink drop in a glass of still water. Initially the ink drop will be well confined, but in time it will spread and after a very long time the water in the entire glass will be almost evenly dyed. This process is called *diffusion*. Our fundamental understanding of diffusion is as follows: both the water as well as the dye in the ink are made of microscopic constituents, which are in constant motion. At macroscopic scales this motion is manifested by what we know as the *temperature* of the liquid. The motion of the constituents is random, some will move up, others will move down, some left others right, some forward and others backward, but the net result, when we consider a large number of constituents, is that there is no net motion in any one direction, so that the whole appears static despite the constant microscopic motion. If we place an ink drop in a glass of still water, then due to the microscopic motion there will be a penetration of the ink drop by water molecules and the dye particles will move into the surrounding water. There is practically no way of describing the microscopic motion mathematically, because the number of constituents (ink and water molecules) is simply too large; besides it is practically impossible to measure all the positions and momenta of the constituents with the necessary accuracy. But there is a way to describe the diffusion process effectively by considering the density of dye particles in a small volume of water. We can now calculate how this density changes without knowing the precise microscopic motion. To do this we calculate the probability of the microscopic motion of the dye particles outside the small volume to move inside this volume and subtract the probability of dye particles inside the volume to move outside. This yields the net change of dye particles in this volume which tells us how the density of dye particles in the small volume changes. To calculate the probability of a density change, we have to sum over the individual probabilities of all processes that lead to this change. It turns out that processes for which the action is very small are likely to occur, while processes whose action is larger will practically never occur.

If we now want to answer a correlation question such as "What is the probability of measuring A given that we have measured B?" then we have to sum over all processes that transform a system that has property B into a system that has property A. It turns out that all possible correlation functions can be bundled together in a mathematical quantity called the *generating functional* that can be calculated in an analogous way. An independent way of thinking about these processes is that they represent a history of the system that is compatible with a certain correlation.

There is, however, a complication: Already for the simple system of a glass of water it is practically impossible to calculate the generating functional because of the huge number of constituents involved. We can only look at very small systems and perform the sum. The reason why this is sufficient is the following very important observation: There is a systematic way of splitting the generating functional into three parts: A relevant part, that becomes larger as the system increases, an irrelevant part that goes to zero as the system increases and a marginal part that stays about the same. It turns out that the mathematical description of the relevant and marginal terms is very simple; the bulk turns out to be irrelevant. We can thus focus on the relevant and marginal parts of the generating functional by performing the sum for a system with a small number of constituents. Subsequently we have to find a way to incorporate the change in the relevant and marginal parts as the system increases. This change can be calculated with the renormalization group equation, which I will introduce in more detail later. A very rough description is that it is an equation that tells us how the generating functional changes when we increase the size of the system a tiny bit. What is relevant, marginal or irrelevant is also determined by it. The important point that I want to make here is that the macroscopic behavior of the statistical system can be determined by a small system whose relevant and marginal parts of the generating functional determine the behavior of the macroscopic system.

The description of quantum systems is very analogous: It is again possible to bundle the correlation functions into a generating functional. This description is so efficient, that one can use it to define the entire quantum system. There is also a renormalization group equation for it and we can again split the generating functional into relevant, marginal and irrelevant parts. There is, however, an important difference: In the statistical system there is a well understood small system whose generating functional can be calculated explicitly; this serves as the so called "initial condition" for the renormalization group equation[1]. In most approaches to quantum field theory one replaces the initial condition with a non-interacting field theory. This is a boring theory where nothing ever happens: Given N particles with their respective momenta at an initial time, we will find precisely these N particles with their respective momenta at any later time. This is certainly not even close to a good approximation of the real world with its richness of phenomena. Despite this disadvantage one has to acknowledge the fact that we have learned very much about quantum field theories by taking this non-interacting initial condition: Most of our knowledge about quantum field theories comes from perturbation theory, which is based on the non-interacting initial condition.

With these definitions in mind, we can now state the problem of quantum gravity : We saw that gravity is a diffeomorphism-invariant geometric theory. It turns out that there is no diffeomorphism-invariant non-interacting theory of geometry that we could use as an initial condition. What one can do instead is to split the geometry into a background geometry and small peturbations around it; then due to this split there appears a non-interacting diffeomorphism-invariant theory of geometry. However, there are no relevant or marginal diffeomorphism-invariant and physically acceptable terms in the generating functional that describe the small peturbations of the background. One is thus stuck with the boring theory of small peturbations of geometry that cannot explain the vastness of gravitational phenomena, such as the evolution of the universe or the structure of black holes. This problem can also be stated in a different way: We could say that gravity is a diffeomorphism-invariant theory of geometry, so its generating functional has to be a diffeomorphism-invariant functional of geometry and forget about the renormalization group. But without the renormalization group telling us which parts of the generating functional are irrelevant, we are forced to make the most general assumptions. This, however, means that there is an infinite number of unknown parameters in the theory, such that the theory is not predictive, because one needs to perform an infinite number of measurements to find the values of these parameters. This is known as the problem of *nonrenormalizability*.

There are at least two possible solutions to the problem of quantum gravity. The first is a bottom-up approach that tries to construct a quantum system as an initial condition for the renormalization group equation by taking the

---

[1] e.g. the initial condition in the ink drop example is the small microscopic system consisting of water molecules and dye particles.

diffeomorphism symmetry of gravity and the mathematical foundations of quantum mechanics seriously. The most advanced of these theories is loop quantum gravity, which I will introduce in the next section. Following that I will explain the top-down approach, which is based on the observation that under certain circumstances it is possible to use the renormalization group equation itself to find the appropriate initial condition for a quantum field theory. Both approaches currently suffer from shortcomings and I will subsequently argue that a suitable combination of both may be a solution to the problem of quantum gravity.

In short:

- Gravity is a theory of the geometry of spacetime.
- The large scale behavior of quantum systems is determined by an initial system and the renormalization group.
- The problem of quantum gravity is that there is no known fundamental initial system and renormalization group that leads to large scale gravity as we know it.

BOTTOM-UP: LOOP QUANTUM GRAVITY

In the previous section I gave a cartoon picture of quantum theories, but I omitted a precise definition of quantum systems. Such a precise mathematical definition exists and it is closely related to what physicists call *canonical quantization*, but more detail would go beyond the scope of this article. I will rather only present the essential questions that one has to answer.

I stated that the correlation functions describe a quantum theory. These are numbers that encode the probability of finding a certain outcome of a measurement *A* given that another measurement had the outcome *B*. We experimentally measure the observable quantities in a system, which are the building blocks of the correlation functions. This may seem to be a tautology, but the question about the observables of a system is in general very nontrivial, because the mathematical description of a system is very often redundant. We saw that we can deal with this redundancy if we are aware of the symmetries of the system that are related to it. A large constituent of the canonical quantization formalism is the formulation of a precise mathematical description of the observables of a system and a precise mathematical description of the symmetries of the system. This analysis in complicated by the fact that the simplest description of a quantum system requires its formulation at one instant in time, whereas the description of observables and symmetries is most efficient when it is formulated in *all* of spacetime. A description of a physical system at one instant in time is called the canonical description; which is where the "canonical" in "canonical quantization" comes from.

It turns out that gravity can be cast as a canonical system, despite its invariance under general spacetime diffeomorphism transformations that mix different instants in time. This can be done because a small spacetime diffeomorphism transformation can be split into a spatial part that acts as a relabeling of the spatial geometry at an instant in time and a deformation of the so called *time slice*.

The particular formulation of general relativity that underlies loop quantum gravity uses the so-called *Ashtekar variables*, which are mathematical objects that encode geometry and curvature. It turns out that these variables are well adapted to the description of geometry on a lattice. To be more precise it turns out that the loop quantum gravity description of spatial geometry can be paraphrased as a finely woven lattice whose nodes carry quanta of volume and whose edges carry quanta of area. These quanta are the fundamental observables of nature if loop quantum gravity is indeed a correct description of the universe. However, estimates for the size of the volume and area quanta is so small that even with our best microscopes, i.e. the latest generation of particle accelerators, like the Large Hadron Collider in Switzerland, there is no chance of directly detecting them, meaning that as of now loop quantum gravity is an untested speculation just like string theory.

The symmetries of gravity add two interesting twists to this picture: Consider those diffeomorphism transformations that affect only the spatial geometry. Suppose we started with one finely woven lattice as our description of quantum geometry, then a diffeomorphism transformation will deform this lattice into a new lattice. But these transformations are symmetries, so the new lattice must be a description of quantum geometry that is as good as the original, so we have to include the new lattice in our description of quantum geometry as well. As a result, it turns out that this symmetry poses nontrivial conditions on quantum geometry, which we have to describe in terms of *all* possible lattices. Amazingly, there exists a mathematical description of quantum geometry that can be formulated on all lattices in a consistent way. It constitutes the core of loop quantum gravity. Before I discuss the consequences of this, I will briefly mention those diffeomorphism transformations that mix different time slices. Let us in particular assume one instant in time and another one a moment later, then there exists a diffeomorphism transformation that transforms the initial time slice into the later one. But this is a symmetry transformation as well as a time evolution, so we cannot disentangle time evolution from the redundancy in the description of quantum geometry. This is known as the problem of time in quantum gravity.

Let us now return to the consequences of the necessity to describe quantum geometry on all possible lattices. At first we notice that if geometry is described on a lattice then the question: "What is outside the lattice?" is unphysical, because we cannot perform any experiment outside the lattice. To illustrate this suppose that we lived in a two dimensional world whose quantum geometry is described by the lattice printed on a graph paper. If we were particles on this sheet-world then we would only be able to hop from one node to a neighboring one, so at each point in time we would be able to go either left, right back or front, but we would not be able to hop anywhere in between. Let us now add quanta of area to each node and quanta of length to each link then it may be that two nodes appear close to one another on the graph paper, but there may be lots of quanta of length in between them, so they are actually far apart. Similarly one may consider two points on the graph paper that are seemingly far apart, but it may be that there are only a few quanta of length between them, so they are actually close together. This teaches us that the geometry that emerges from quantum geometry may be very different from the geometry of the sheet of paper that was used to draw the graph.

However, there is an even stranger phenomenon: I stated that we have to admit any graph in our description of quantum geometry, so the description of quantum geometry automatically contains graphs that have shortcuts, i.e. we may have a link that connects a node with another node that is far away on the paper. If we are now a quantum particle on such a graph and we arrive at such a node then we can go left, right, backwards, forwards and in addition we can also use the shortcut! In effect it will appear to us as if the far node is just as close as the neighboring ones[2]. One might even think that there is an extra dimension at this node that allows us to go in the direction of the shortcut. Let us now look at this shortcut phenomenon from a different perspective: Suppose we zoom out of the graph, such that the small scale quantum "bumpiness" appears smooth. Then the shortcut appears as a pair of points in this geometry that carry additional quantum numbers. This is very similar to particle-antiparticle pairs, and such an interpretation has been suggested as a possible consequence of loop quantum gravity. This may be rephrased as follows: suppose almost all of the quantized spacetime looked smooth when one zooms out, except for a few isolated defects in the graph. Then these defects may be interpreted as *particles that live* on the geometry. If we believe in such an interpretation then there is a large spectrum of possible particles, because one can build many different kinds of defects, ones with knots, braids, different quantum numbers and so on (describing different features of the particles like mass, charge etc). I want to caution here that even if loop quantum gravity turns out to be correct then we would still have to check whether these defects actually have the properties of the observed particles, in particular their dynamics. This is, however, a question whose answer is currently inaccessible within the framework of loop quantum gravity.

All of this leads to the major construction effort of loop quantum gravity: the precise implementation of those diffeomorphisms. There are proposals, most prominently the so-called *spin foam models* and *group field theories*, but it is so far unclear as to what extent these models describe nature. The basic problem is that one has to show how ordinary gravity arises in these models, which so far can only be done within certain approximations and the quality of these approximations is not well-understood. This is mainly because we cannot really say what "zooming out" means in the language of loop quantum gravity. On the one hand the large scale quantum geometry may be very different from the geometry of the graph and on the other even the graph itself may have defects that are not compatible with large scale geometry. Both effects lead to problems if one tries to give a prescription of what zooming out should mean in loop quantum gravity. But without being able to zoom out, we are unable to address what the relevant and marginal components of the theory are and we are thus unable to answer whether the

---

[2] *Editor's note*: As in figure 11 on page 74

microscopic theory that we constructed can really reproduce the richness of the gravitational phenomena that we observe in nature.

In short:

- Loop quantum gravity is a mathematically well defined theory that predicts a discrete microscopic structure rich enough to describe gravity in terms of quantum geometry. This structure, however, seems to contain more than just pure geometry.
- Problem: It is unknown how large scale gravity as we know it could emerge from the microscopic theory.

TOP-DOWN: ASYMPTOTIC SAFETY

Asymptotic safety is a particular scenario whose conditions, if met, allow one to define a quantum field theory without an initial condition for the renormalization group equation. It has been pointed out by Nobel laureate Steven Weinberg that this could be a resolution to the problems of quantum gravity. To explain this scenario I will give a more detailed explanation of the renormalization group equation. Thereafter I will review the present day evidence in favor of this scenario and discuss the open questions.

THE RENORMALIZATION GROUP REVISITED

To get a closer look at the renormalization group equation, consider the following example: The description of a macroscopic body, say a soccer ball, in terms of its constituent elementary particles is a hopelessly difficult task and, for most practical issues, quite unnecessary. Physicists are trained to find practical effective descriptions for complicated systems; e.g. for predicting the trajectory of a soccer ball they may approximate the ball by a rigid body and derive very accurate results despite ignoring “almost all” of the ball’s microscopic description. The recipe for achieving a good effective description is:

1. Finding the appropriate “zoom level” (scale) for the question asked and
2. Parameterizing our ignorance of everything “smaller than” the zoom level as properties of the effective model.

By doing this we have to expect that the effective laws of physics change with the zoom level. One way of introducing the renormalization group equation is to define it as the equation that tells us how the laws of physics change when we are zooming.

To explain this I have to disentangle some terminology: I explained that quantum theories can be encoded using correlation functions and that these can be conveniently bundled into a mathematical object called a generating functional. There is, however, an even more compact way to encode a quantum field theory, and that is by describing it using an object called the *effective action.* For our purposes, it is not important to understand its mathematical definition; suffice it to say that the resulting effective action is usually an object of a much simpler structure than the generating functional. Moreover, the name “effective action” suggests two properties: The first is that it plays the role of an action for the quantum theory, in the sense that the equations calculated form it encode the entire quantum system. The other property is that the effective action takes the same form as a classical action. This is very important because there is a natural way to organize the possible terms in the action. This organizing principle is due to locality ; meaning that the most important interactions are at the same point or between points very close to one another while the more distant interactions become less and less important. The strengths of these interactions are numerical constants known as the coupling constants of the quantum field theory. In principle there is an infinity of interactions and hence there is an infinity of coupling constants needed to fully specify the effective action. However, the principle of locality (together with dimensional analysis[3]) provides us with an organizing principle for these constants that tells us which of them may be important.

[3] Although dimensional analysis is very important for the organization of couplings, I will not discuss it here, because it goes beyond the scope of this article.

To obtain the effective action at a finite zoom level, let us now recall how we preformed the zooming out in the ink drop example: Instead of looking at the motion of all dye particles and all water molecules, we considered the average number of dye particles in a small volume. We thus coarse-grained the model and kept track of only the average density of dye particles in the volume. The same thing can be done for the effective action of a quantum field theory and the object obtained from such a coarse-graining is called the *effective average action*. It turns out that it is often possible to organize the effective average action according to locality. The change of the laws of physics can now be quantified as the change of the effective average action when we change the zoom level by changing the coarse-grained volume. The equation for the change in the effective average action can be translated into an equation for the change of the coupling constants, so it tells us how the strength of the different interactions changes through zooming. It further turns out that the renormalization group equation that describes this change takes a very simple form, which is the reason why it is so useful.

I will now sketch how this renormalization group equation is formulated: I explained that the generating functional can be expressed as a sum over histories. This sum is called the *path integral* and the quantity that is summed over can be expressed in terms of a quantity called the *bare action*. Neither the path integral nor the bare action have a rigorous mathematical definition, but the idea of summing a quantity that has the properties of an action over all histories is a very good insight that helps mathematical physicists find mathematically well-defined objects[4]. The basic complication is that there are usually too many histories, so the sum over all histories contains uncountably many terms. One way of obtaining a rigorous definition of a path integral is to restrict the sum over all histories to some subset of "most important" histories. Physicists call this procedure a *cut-off*, because the sum over all histories is reduced to a piece that contains the most important ones and the rest are cut off from the sum. The precise implementation of a cut-off is tricky but once it is properly done one is able to define both a path integral and a bare action. Almost all cut-offs are based on the idea that one removes histories that lead only to very short distance effects. In which case there is a good physical interpretation of the bare action that leads to a fundamental insight: We may think of the bare action as an object that is obtained by performing the sum over short distance histories. In other words: we can think of the path integral with cut-off as a path integral that is already partially performed and that the effect of the histories that are cut off is expressed through the bare action. It turns out that once a path integral is constructed one is sometimes able to remove the cut-off. The bare action in the theory without cut-off may then be seen as the *fundamental bare action*. I want to caution the reader at this point that many authors use the term "bare action" to describe what I am calling the "fundamental bare action" and they use the term "effective action at the cut-off" for what I call the "bare action".

Let us now assume we have a path integral with a cut-off and a bare action as described here, for which we want to derive a renormalization group equation. The question is: How can we implement the zooming out or coarse-graining procedure? The answer lies in the observation that was made in the previous paragraph: We saw that the path integral with cut-off may be thought of as a path integral without a cut-off where we already performed the sum over those histories that affect only short distance physics. Let us now consider a second path integral with a cut-off that is at slightly larger distances. We can think of this second path integral as the first one plus a little more summation performed. But the summation over this small number can be explicitly done, so we can exactly calculate how the bare action of the first path integral changes into the one in the second path integral. We now call the action in the second path integral the *Wilsonian action* or *Wilsonian effective action*, because this idea is to a large extent due to Nobel laureate Kenneth Wilson. If the change in the cut-off is very small then one can use the path integral to derive an exact equation that tells us how the Wilsonian action changes as the cut-off is changed. This exact equation is the renormalization group equation. An analogous procedure can be performed to derive the equation for the effective average action. In this case the equation is called the *functional renormalization group equation*, which tells us how the effective average action changes as we change the coarse-graining.

I stated before that there is a natural organizing principle for the coupling constants (i.e. the strengths of the interactions) in the effective average action and that the functional renormalization group equation can be translated into an equation that tells us how these constants change as we change the zoom level. The equations in this case are known as the *beta functions* of the coupling constants. We can now inspect these functions and determine whether a coupling constant becomes stronger, weaker or stays about the same as we zoom out. In analogy to the ink drop

---

[4] The quantity that is summed over is actually the exponential of the action, not the action itself.

statistical system, we call coupling constants that become stronger as we zoom out *relevant*, the ones that become weaker *irrelevant* and the remaining ones marginal.

Up until now we have been conservative in accepting the cut-off and justifying the use of a bare action as a parameterization of what we do not know about the theory. This is not just a mathematical question, but it is also very physical, because it makes the theory predictive. This can be seen as follows: If we consider a path integral with cut-off then we can introduce an infinite number of coupling constants into the bare action and thus need an infinite number of measurements to determine the bare action and thus we can never use the theory to make predictions. On the other hand the existence conditions for a fundamental bare action in a path integral without cut-off are much more stringent and may indeed lead to predictive theories. We can therefore be bold and ask for a sufficient condition that allows us to remove the cut-off from the path integral and thus pretend that we have a theory at *all* zoom levels. The question then is to ask whether there exists a fundamental bare action for the system. We can find this because the functional renormalization group equation can also be used in "reverse gear"; this means that we can also use it to calculate how the bare action has to change when we zoom in rather than out. It very often happens that the renormalization group tells us that some coupling constants become infinite as we zoom in. A theory that needs such couplings does not make sense at arbitrarily fine zoom levels. However, sometimes we do not run into this problem and we can make sense of the theory at any zoom level.

The simplest way in which this may be realized is if the couplings do not change or, equivalently, that all beta functions vanish, meaning that we have found a so-called *fixed point* for the renormalization group equation. Physically acceptable fixed points are sparse, and this is the reason why there are not many quantum field theories that are believed to exist. After a fixed point is found, one may consider which values of the coupling constants go into the fixed point as we zoom in. This is an important question, because when we perform experiments we are probing the quantum field theory at a finite zoom level and we will thus not measure the coupling constants of the fundamental bare action, but rather the effective coupling constants at the experiment's zoom level. If we consider the coupling constants as coordinates in a hypothetical space that contains all effective actions then the fixed point is one point in this space, but those values of the coupling constants that move into the fixed point as we increase the zoom level usually form a surface in this space. This surface is called the *critical surface*.

Let us explain this using an example: It turns out that fixed points exist when all interactions are turned off. These, however, define a boring theory where nothing happens. However if there is (at least) one coupling constant that becomes stronger as we zoom out, then an interacting theory will emerge. This scenario is realized in the theory of the strong nuclear interactions, which contains a rich particle spectrum of mesons, hadrons and nuclei at large distances, but if we zoom in we see that the interactions between the more elementary quarks and gluons that are the building blocks of the mesons, hadrons and nuclei become weaker and weaker. This phenomenon is called *asymptotic freedom*, because the quarks and gluons become free non-interacting particles if we zoom in at arbitrarily fine zoom levels. We conclude that the theory of strong interactions makes sense at arbitrary zoom levels. In addition it turns out to be predictive as well.

In short:

- The renormalization group tells us how the laws of physics change as we zoom out of a system.
- Some theories allow zooming in and out to any level, and may be seen as fundamental. These theories are fixed points of the renormalization group.

THE ASYMPTOTIC SAFETY SCENARIO

We have just seen that a theory with a non-interacting fixed point and a small number of relevant coupling constants allows for the rigorous definition of a quantum theory. We can now consider a more general scenario. We want the theory to be fundamental, so we cannot use a theory with a cut-off, but have to ask for a theory that makes sense at all zoom levels. We saw that the fixed points of the renormalization group equation tell us what kind of theories can satisfy this criterion. But there is still one caveat: Even if we found a fixed point of the renormalization group, there may exist an infinite number of independent coupling constants that will go into the fixed point as we increase the zoom level. Then, although the theory makes sense at all levels, it is still not predictive, because we will again have to perform an infinite number of measurements to determine the fundamental bare action. So, if we want to be able to determine the fundamental bare action from a finite number of experimental data then we have to demand that there is only a finite number of independent coupling constants that move into the fixed point as we zoom in. One says that we demand that the "critical surface is finite dimensional".

We may now consider a general fixed point of the renormalization group, i.e. one that is not necessarily a non-interacting free field theory. It will only be predictive if its critical surface is finite dimensional. Such a quantum field theory is called *asymptotically safe*. Asymptotically free theories are a special case of asymptotically safe theories, but they are not the only ones.

The cut-off can be removed from asymptotically safe theories and we can hence view them as fundamental, because we did not have to partially perform the path integral to obtain a bare action at the cut-off. But even if we find an acceptable theory that exhibits asymptotic safety, one cannot be sure that this theory will describe nature at all the zoom levels that have not yet been experimentally tested. This is because measurements will always be limited to a finite accuracy, but for a theory to surely be asymptotically safe all of its coupling constants have to lie within the critical surface with no margin of error, which would require an infinite number of measurements with absolute precision. So is there any value of finding an acceptable asymptotically safe theory if it does not imply that this theory is fundamental? The answer is in the affirmative: Suppose that there is a small scale theory whose path integral is partially performed to yield a theory with a cut-off at a finite distance scale far smaller than our current experimentally accessible scales. This theory will in general have all kinds of coupling constants. But if we now use the renormalization group equation to zoom out towards our experimentally accessible distance scale then all irrelevant constants will become weaker and weaker and essentially die out. We thus expect that the further we zoom in the closer the theory will get to the critical surface. This phenomenon is known as *universality* : It means that the long distance physics is often independent of the short distance behavior. Moreover, if the critical surface is finite dimensional, i.e. there is only a finite number of independent coupling constants, then we are able to make stable predictions for long distance physics without ever having to bother about a short distance theory. On the other hand, if we observe significant deviations from the critical surface in our experiments, then we know that we are close to the cut-off of our theory and we can thus predict that there has to be "new physics" beyond this cut-off.

In short:

- Fixed points of the renormalization group are sparse.
- Some fixed point theories only have a finite number of relevant coupling constants and are thus determined with a finite number of measurements. These theories are called asymptotically safe.

THE ASYMPTOTIC SAFETY SCENARIO IN QUANTUM GRAVITY

There has been a recent surge in activity towards the question of whether quantum gravity may be asymptotically safe. Research in this direction is very technical and I will not explain the efforts in detail, but rather focus on the setup, the preliminary results and some open questions.

One main problem in the construction of a renormalization group equation for quantum gravity is the definition of what is the required zoom level. Recall that a theory of gravity is in fact a theory of the geometry of spacetime itself and thus tells us how "distance" is defined. This leads to a vicious cycle: on the one hand we need gravity to tell us what "distance" is, or in other words without having gravity we do not know what "distance" is, on the other we need a notion of distance so we can tell how to zoom out, so we can say what gravity is in the context of the renormalization group. The cycle can be broken with a trick: For pure convenience we introduce two types of gravity; "gravity A" and "gravity B". Then we use gravity A to define "distance" (and hence zooming) for gravity B and such that we can set up a path integral with cut-off for gravity B in the presence of gravity A. We then use this setup to derive a functional renormalization group equation for the combined system of gravities A and B. At the end of the calculation we remember that there is only one kind of gravity so we set gravity A equal to gravity B. This trick is known as the *background field method*, although this name is an unfortunate misnomer, because it is precisely through this method that it becomes possible to derive a renormalization group equation for gravity that does *not* depend on any artificially introduced background.

Having a functional renormalization group equation at our disposal, we can in principle test the asymptotic safety scenario. The equations however turn out to be very complicated, so one is practically forced to use approximations in the sense that one includes only a small number of coupling constants in the effective average action. These approximations are called *truncations*, because they amount to truncating the effective average action. By now various truncations for quantum gravity have been considered and all of these exhibit the same picture: there appears to be a nontrivial fixed point for the functional renormalization group that has a three dimensional critical surface, i.e. the theory appears to have three independent coupling constants, compatible with asymptotic safety. Moreover, it turns out that the quality of the fixed point remains the same in all the truncations that have been considered so far. This is important; because one of the main objections to these investigations is that the fixed point may be an artifact of the truncations. But if one can go to larger and larger truncations without changing the quality of the fixed point then it becomes less and less likely that the fixed point is such an artifact.

Despite these encouraging results, there are still many open questions. The most important of course is whether gravity really is asymptotically safe? So far there is no proof of this either way. The existence of a fixed point is usually due to a special symmetry that manifests itself at the fixed point. So the question may be rephrased to whether one could identify the symmetry that governs the fixed point. And even if one could establish the existence of the fixed point, one would still have to work hard to calculate predictions from this scenario.

In short:

- It is practically very difficult to check the asymptotic safety conditions, so approximations have to be used.
- All approximations used so far point towards the existence of an asymptotically safe quantum gravity theory, but there is no conclusive proof.
- The main problem is that we still do not have a good guess of what the fixed point looks like.

THE MEETING POINT

A subtitle for this section could be "A renormalization group equation for loop quantum gravity" or "Loop quantum gravity as an initial condition for the renormalization group". This dichotomy of subtitles summarizes my vision that both approaches to quantum gravity may really be complementary. We saw on the one hand that the loop quantum gravity framework is very well adapted to an implementation of quantum theory in the gravitational framework, but that it is so far impossible to check whether this theory can indeed be used to describe gravity at large scales. This problem is trivial in the renormalization group approach to quantum gravity, but it is technically very difficult to establish a fixed point that can serve as a theory of quantum gravity at all zoom levels. We see that the strength of one theory is met by a weakness of the other and vice versa. It is therefore very natural to assume that significant progress for both approaches can be made if the methods of the other approach can be applied. In the best of all scenarios, one would arrive at a completely unified picture that contains a particular version of loop quantum gravity as a fixed point for the renormalization group. I will now envision such a scenario and discuss its possible consequences.

The fundamental ingredient of the renormalization group is the introduction of the concept of scale, or zoom level. We saw that the trick by which this is done was the use of two kinds of gravity, which I called gravity *A* and *B* in the previous section. If we want to implement the same trick in the loop quantum gravity framework then we should implement a similar notion of twofold gravities or twofold geometries. Then we could define the zoom level again with respect to one of these, say gravity *A*. We remember that gravity is built from discrete blocks, but after a zoom level is defined, we might do what we have done in the ink drop example, where a discrete number of water molecules and dye particles played the role of the fundamental constituents, but after zooming out we were able to approximate the distribution of ink particles by a smooth average density in a small volume. We might therefore be able to formulate the renormalization group in terms of two smooth geometries, but it is likely that the renormalization group equation gets modified, because we are using a different setup. Let me explain how these modifications might arise.

We saw that loop quantum gravity contained not just smooth geometry, but inherently also required small scale defects. We argued that these defects may have the properties of particles, and if we want to incorporate their effects in a setup with smooth geometry then we indeed have to treat them as separate fields: Let us for simplicity only consider one kind of defect and assume that the rest of the quantum geometry is well approximated by a smooth geometry. If we zoom out, we would not see individual defects anymore but only an average density of such defects within a small volume, which can be described by a smooth field, just as in the case of the density of molecules in water. However, there is a zoology of possible defects, meaning that loop quantum gravity predicts its own particular "extra fields" (or extra particles) content. This is interesting in two ways: On the one hand I explained that the existence of a fixed point may be explained by the existence of a special symmetry that governs it. Having an extra particle content of the type predicted by loop quantum gravity, one can search for possible symmetries that mix between spacetime and the particle content and this symmetry may have the power to explain the fixed point. On the other hand, there is the interesting possibility that part of the extra particle content may be what we experience as matter; if so then loop quantum gravity may be a theory of pure gravity that is evolving into a theory of everything. However, to be able to identify the extra particles content as physical particles one has to make sure that their large scale dynamics and interactions exactly match those that are observed in nature. This is nontrivial because the renormalization group may predict that some, or maybe even most, extra fields are non-dynamical at large scales; their dynamics and interactions may all be irrelevant, or may have the wrong large scale interactions, i.e. the relevant interactions do not match the ones observed in nature.

There are, however, a few other questions that a merger of loop quantum gravity and the renormalization group may be able to shed light upon. I will highlight only one example: Loop quantum gravity predicts a discrete geometry, in particular a discrete area spectrum with a smallest possible area, and requires that area measurements of two intersecting areas influence one another. Such effects also occur in a mathematical setting called *noncommutative geometry*[5]. This is a mathematical framework that can be used to describe normal geometry, but its strength is that it is much more general and can be used to describe very interesting structures whose nature is not at all geometric; nevertheless one can still apply many of its useful geometric concepts. In using the rules of noncommutative geometry it is actually possible to describe particles that interact with geometry as being part of the geometry itself, but there are conditions that have to be satisfied for this to work. Remarkably, it turns out that the particle content we see in nature, the Standard Model of particle physics, and its action fulfills these conditions in a nontrivial way. This noncommutative framework is called the *spectral action*, and it turns out that the mathematical tools that are used to derive the Standard Model from noncommutative geometry are the same tools that are used to perform calculations in the renormalization group framework, so one can hope that the two frameworks may be combined one day. Moreover, it turns out that the spectral action framework in noncommutative geometry predicts an action at a high energy scale and that it is necessary to use the renormalization group to derive what its constraints predict for the effective action at current experimental scales. If we now assume that loop quantum gravity is correct then the particles and the initial action will follow from it, and it is not unreasonable to assume that if loop quantum gravity is right it will predict a particular noncommutative geometry. The renormalization group will then tell us how the transition from the small scale structure provided by loop quantum gravity, having properties of a noncommutative nature, to the large scale geometry as we know it, and can be viewed as a zooming process. This would yield groundbreaking insights that would change the understanding of our world. Such insights would be of a mathematical nature and thus further add to the tool kit of theoretical physicists. These new tools would be valuable, even if loop quantum gravity or the asymptotic safety scenario were both proven wrong.

We can, however, also assume an opposite scenario, namely that it would first be established that the renormalization group has a fixed point with all the properties that are required to describe nature. It is likely that a mathematical result of this kind would not be constructive, meaning that we would know for sure that the fixed point exists and has certain properties, but we would lack an explicit description. While such a result would be a major breakthrough, it would not help us in working out experimental predictions, so we would still have to look for a framework that tells us what the fixed point looks like. Since loop quantum gravity is very well adapted to the symmetries of gravity, it is likely that its framework would be useful for the description of the fixed point, and it may turn out that a particular version of loop quantum gravity indeed serves as the initial condition for the renormalization group.

But even in the case where one is able to prove the existence of a physically acceptable fixed point there is still the possibility that loop quantum gravity acts as the initial condition. In this case however, one would be able to derive

[5] *Editor's note*: See "*Science Fiction of Everything*".

loop quantum gravity and in particular its implementation of the symmetries of gravity from the fixed point and its symmetries.

A first hint for such a scenario may already have been found: The renormalization group predicts that the geometry of spacetime becomes effectively two-dimensional at the fixed point and this is compatible with the observation that the building blocks of loop quantum gravity's spatial geometry are one-dimensional loops. I leave it to the reader to judge whether this is a coincidence or whether it serves as a motivation for my speculations.

I want to conclude with again admitting the speculative nature of finding a relation between loop quantum gravity and the renormalization group/asymptotic safety framework, and I want to caution that my speculations in this article go one step beyond the speculative nature of the current approaches to quantum gravity, because I speculate not only about what might be realized in nature, but also how the mathematics that is supposed to describe it *could* work out. But this is where I actually see the most potential, because as long as we have not worked out all the details we run into the possibility of finding a surprise at the next step. History has shown us that the biggest surprises happen when they are least expected; and may already be happening right now to someone working on a completely different idea.

CHAPTER 7

Does History Repeat Itself in a Non-Cyclic Universe?

Sundance Bilson-Thompson

School of Chemistry and Physics, University of Adelaide, SA 5005, Australia

Abstract: When trying to construct a working theory of quantum gravity, it may be helpful to consider how previous breakthroughs in physics have occurred. What can we learn from the historical progress of astrophysics and particle physics? What role should theory and experiment play as we try to find our way towards this seemingly elusive goal?

INTRODUCTION

It's tempting to think of a theory of quantum gravity as some sort of ultimate goal of modern physics. We have two fundamental theories (the Standard Model and general relativity), each essentially uncontested in its own realm of validity. If we could combine them harmoniously we would, hopefully, have a single theory encompassing every force and every length scale. And naturally we would hope that this theory would provide new insights into the mysteries that seem to challenge our modern view of the universe.

I suspect, however, that instead, a successful theory of quantum gravity will startle us with how obvious it looks in retrospect. And it will bring us to a better appreciation of the assumptions that unwittingly go into our physical theories. I won't pretend for a moment to have a definite idea of what such a theory will look like (although I have my suspicions), but if previous scientific watersheds are anything to go by it will have two major characteristics; it will be prompted by experimental or observational results, and it will require us to abandon some basic assumption about how the world works. The history of the physical sciences is full of examples of such breakthroughs, and they should serve as cautionary tales whenever the hubris to which we theorists are prone leads us to believe that we can understand nature through pure thought alone. Another specific characteristic it must have is the ability to explain prior, well-established observations.

ORBITS AND EPICYCLES

Consider the Keplerian revolution. This was the time when the idea that the Earth was at the center of the universe was replaced by an understanding that the Earth is a planet, and orbits the Sun. Most people call it the Copernican revolution, but I think this places the emphasis on the wrong person. Nicolaus Copernicus proposed a Sun-centered model of the universe, but it was Johannes Kepler, in the early seventeenth century, who turned this into an almost-modern theory of dynamical bodies and forces. In fact, Copernicus got the idea of a Sun-centered universe from the ancient Greeks (most notably Aristarchus of Samos). Until Copernicus' time it was believed that the planets and the other heavenly bodies were "perfect", in the sense that they followed perfectly circular paths, traveling at a uniform rate around the Earth. Inconveniently, the observed motions of the planets suggested that they did not move in such a manner. Copernicus actually proposed the Sun-centered hypothesis as a way of reconciling the observations more closely with the idea of perfect circular motion.

Kepler 's contribution was more profound. He set out to understand why the planetary orbits had the sizes they did. This prompted him to use the observational data obtained by Tycho Brahe (the best astronomical data known before the advent of the telescope) to calculate the exact shape of Mars' orbit. After a monumental mathematical effort (during which he almost invented integral calculus) he discovered his first law of planetary motion; the planets move around the Sun in elliptical (not circular) orbits with the Sun at one focus – see (Fig. **1**). While Kepler discovered two other laws of planetary motion, which are significant for their quantitative nature, it is the first law that was so pivotal in breaking down a long-held assumption. Arguably, Kepler's first law separated astronomy from religion by overthrowing the idea that one set of behaviors was inherent to the "perfect" heavenly bodies (which occupied the realm where divine beings resided), and a different set was inherent to Earthly bodies, including humans. He profoundly altered a basic assumption about the nature of the visible universe, and in his writings excitedly makes a statement – obvious to modern-day readers – that "astronomy is part of physics". Modern theoretical physics deals with laws (e.g. the law of gravitation) that we expect to be valid everywhere throughout the Universe. In that sense, Kepler's work marks the beginnings of modern physics, by introducing the idea of what we now call a *universal law*.

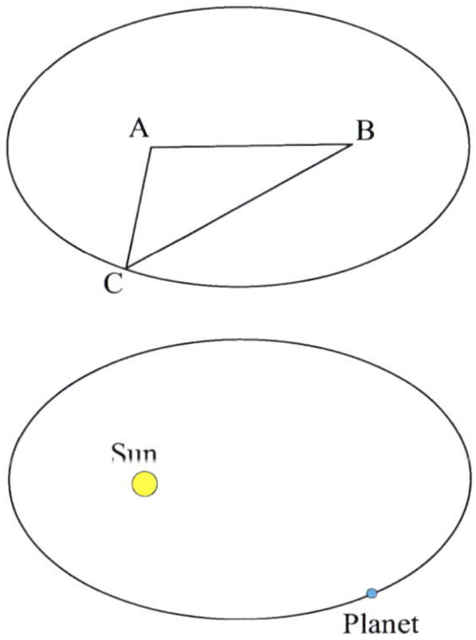

Figure 1: An ellipse has two focus points, A and B, such that the total distance from A to B to C to A is the same, no matter where C is (top). Kepler found that the orbit of a planet is an ellipse, with the Sun at one focus of the ellipse.

It is interesting to note that Kepler systematically abandoned assumptions about the motion of the planets until he was able to solve the problem of finding Mars' orbit. And yet, the elliptical nature of the planetary orbits had been "known" for centuries. Astronomers had proposed the existence of the so-called *epicycles* – sequences of circular motions stacked one upon the other – in a convoluted attempt to preserve the idea that the planets all moved at constant rates along perfectly circular paths – as shown in (Fig. **2**). Nonetheless, the actual motions of the planets through space could be seen to be roughly egg-shaped, once all the epicyclical motions were added together. Kepler simply took the heroic step of accepting observational data *at face value*, and in fact chided himself for not accepting the obvious truth sooner than he actually did.

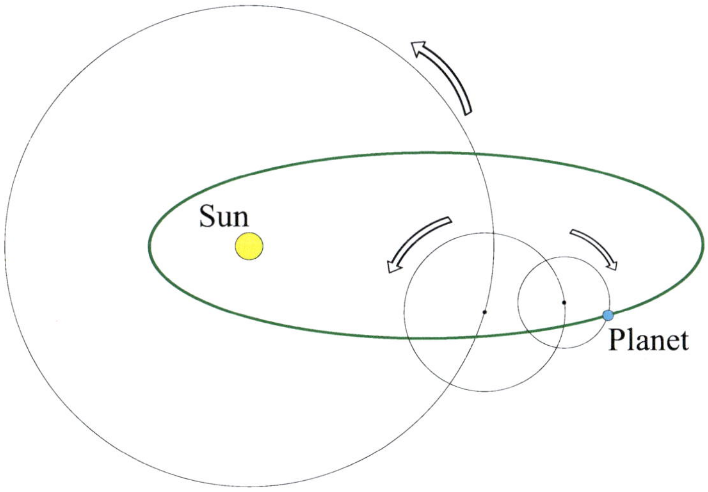

Figure 2: By combining many circular motions (epicycles), each centered upon the edge of another, ancient astronomers were able to reproduce close approximations to the true motion of the planets (green ellipse) using only circular motion.

LIGHTWAVES AND TIMEWARPS

Another example of a scientific watershed was associated with the nature of light. Was it a type of wave, or was it composed of particles? The question seemed to have been resolved by Young 's double-slit experiment (in which light passing through two parallel slits produced a series of bright and dark bands on a screen, signaling a wave-like behavior of light, similar to the way ripples on a pond behave when they pass around obstacles), and by Maxwell 's unification of the electric and magnetic forces into a single phenomenon; electromagnetism. Maxwell began with four equations that had been previously discovered, modifying them slightly to treat electricity and magnetism in a more symmetrical manner. When these four equations were combined they indicated that a self-sustaining wave of electric and magnetic fields could exist, each electric wave producing a magnetic wave, which in turn produced an electric wave, and so on indefinitely. An automatic consequence of Maxwell's equations was that such an electromagnetic wave would travel at the exact speed at which light was known to travel. His inevitable conclusion was that if it looked like an electromagnetic duck, quacked like an electromagnetic duck, and traveled at the same speed as an electromagnetic duck,... well, it was clear that light was a wave.

We now know that light is not just a wave, but displays both wave-like and particle-like properties. But this realization was not arrived at simply on theoretical grounds. It was prompted by the observation that when light shines on a piece of metal, it causes electrons to be "kicked" out of the surface of the metal. This is the so-called *photoelectric effect,* and the strange thing about it is that the intensity of the light used is not as important as its wavelength. For any metal there is a threshold wavelength (or equivalently, a threshold frequency). If the light has a frequency below this threshold it will never kick electrons out of the metal, no matter how bright it is, even though it seems natural to associate the brightness of light with its intensity (or how much "kick" it delivers). This odd effect was explained by Einstein, when he proposed that light comes in little packets (*quanta*), each carrying a certain amount of energy. When light shines on an electron, the quanta collide with the electron. If the electron absorbs a quantum of light, it absorbs *all* the energy of that quantum – no more, no less. Below the threshold frequency, each quantum does not have enough energy to liberate an electron from the metal, and so the quanta are absorbed and re-emitted without kicking the electrons loose from the metal, no matter how many quanta of light hit the metal, i.e. no matter what intensity the light had. The fact that light behaved like discrete particles or packets of energy, yet also produced a wave-like pattern in double-slit experiments forced the concept of wave-particle duality upon early 20th century physicists. This concept eventually extended to the electrons themselves, as well as all the particles of the subatomic world. And it forced us to abandon the assumption that matter and energy had to be either a particle or a wave. They, in fact, have the properties of both!

Maxwell 's deduction that light was an electromagnetic wave raised an intriguing question. If he could calculate the speed of light from his equations, what was that speed measured relative to? The equations seemed to say that the speed of light is not measured relative to anything; it is, in some sense, absolute. Eventually, after people had tried and failed to detect a medium in which light propagated, it was Einstein again who, like Kepler before him, took the bold step of accepting the experimental and observational results at face value, and proposing that the speed of light measured by all non-accelerating observers was the same. This realization came at a price though. Suppose I'm looking at myself in a mirror, and both I and the mirror are moving forward at close to the speed of light. If you watch us zoom past, you'll notice that the light carrying my image to the mirror takes a lot longer than usual to reach the mirror (because the mirror is moving in the same direction as the light, and hence "running away" from it). However I see no such thing, because as far as I'm concerned the light is still moving from my face to the mirror at the speed of light. The only resolution to this paradox must be that all my perceptions and actions are slowed down (as viewed by you), so that I don't notice the light propagating slowly from my face to the mirror. We are forced to conclude that the speed of light is a universal absolute, but time is not – that the rate at which time passes is mutable, and any non-accelerating observer can claim that their measurement of time is just as valid as anyone else's. But speed is distance divided by a time interval. If the speed of light is constant, and time intervals can be measured differently, then distances must also be measured differently by different observers, to compensate for the mutability of time. We are forced to abandon the assumption that time and space are absolute, universal, and unchanging.

WHAT MIGHT WE LEARN FROM THIS?

The middle of the previous century saw a tremendous growth in the area of particle physics. One of the most fascinating discoveries of this period was the existence of multiple "generations" of particles – that is to say that along with electrons and neutrinos there are also similar particles with the same electric charges and other properties, but greater masses. The same is true of the quarks we now know to be the constituents of protons and neutrons. Once again, this was not a theoretical deduction, but came about as a result of observations and experiments, especially the discovery of previously unknown particles in cosmic rays.

It is true that theoretical developments have filled in the gaps between such experimental discoveries. For example, Newton brilliantly developed a mathematical understanding of mechanics, and was able to show that Kepler 's three laws of planetary motion are a consequence of his theoretical framework. Hamilton and Lagrange developed ways of understanding Newton's mechanics in terms of energy, in a manner which would eventually prove central to the formulation of quantum mechanics. Maxwell 's unification of electricity and magnetism was motivated by purely theoretical considerations. Einstein famously developed much of special and general relativity on the basis of thought experiments alone. The staggering profusion of particles discovered in collider experiments was greatly simplified by using ideas from the mathematical field of *group theory*, and these ideas enabled theorists to predict the existence as well as the masses of hitherto undiscovered particles. Theoretical physicists have justly become proud of their ability to predict a phenomenon in advance, and even suggest the experiments needed to confirm or refute these predictions. But the growth of physics, since Kepler's time, has been a mix of theoretical predictions that carry our knowledge of the universe forward, and experimental/observational surprises that act as "course corrections", telling us when we're on the wrong track, forcing us to abandon our cherished assumptions. The perfect circularity of heavenly motion, the universality of space and time, the either-wave-or-particle nature of matter have all been "obviously" true before they were shown to be false. Dealing specifically with quantum gravity, a great deal is often made of the "rivalry" between string theory and loop quantum gravity. And while some researchers have strong views, most of us recognize that we theorists can argue amongst ourselves about who has the best theory until we are blue in the face, but at the end of the day it is experiments and observations that get the final say.

I fully expect that a new round of experimental results will be needed to break the intellectual deadlock that stands between us and a working theory of quantum gravity. And such a theory will, if history has taught us anything, force us to question some basic assumption about the nature of Nature; an assumption that we hold so close we can barely see it. Unfortunately, there are no experiments taking place in the near future which we can confidently expect to break the current deadlock. This sounds like a hopeless situation, but the good news is that the facts we need to find in order to accomplish this breakthrough, like the elliptical nature of planetary orbits, may already be known to us, and are simply awaiting interpretation. As I said earlier, I won't pretend to know what a complete theory of quantum gravity will look like, but there are some points I would like to chew on which may light the way.

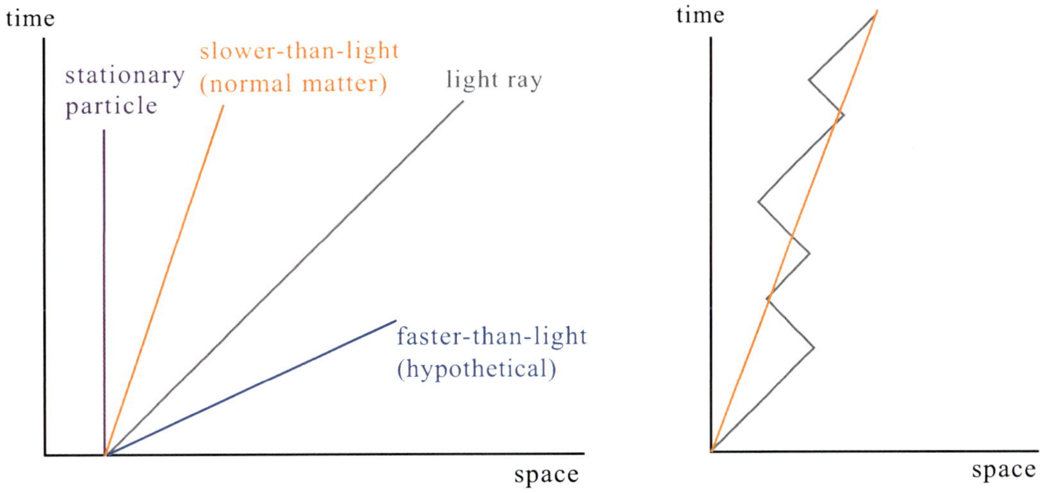

Figure 3: Particles moving at different speeds move from bottom (present) to top (future) trace out different paths on a spacetime diagram (left). A particle moving slower than light can be thought of as zig-zagging back and forth at the speed of light (right).

Gravity is best described by Einstein 's general theory of relativity. The behavior of the subatomic world is described by quantum field theory. General relativity is what we call a classical theory (one in which all quantities such as energy and position are perfectly well-defined, at least in principle). On the other hand, quantum field theory, as the name suggests, is a quantum theory, where things are not as well defined, not in the same sense anyway. In general relativity, when a system goes from one state to another (say, a planet orbiting from one side of a star to the other) it does so in just one way. The gravitational field of a planet or a star is described as the curvature of spacetime in the region surrounding it. In quantum field theory, when a system goes from one state to another (say, an electron propagating from one side of a perforated screen to the other) it does so in every possible way, and the observed outcome is the result of all those different ways "added together". The field surrounding a charged particle, such as the electric field around a proton, is a cloud of "messenger" particles (called bosons) which couple to (interact with) other particles with compatible charges in the vicinity. Sometimes the interaction of a charged particle with a boson turns it into a different type of particle. For example, an electron may interact with a type of particle known as a W^+ boson, and turn from an electron into a neutrino (neutrinos don't carry electric charge, but they and electrons do carry a different type of charge which allows them to interact via the Weak force). Alternatively the interaction could go the other way, with a neutrino turning into an electron. We say that the Weak force *couples* electrons and neutrinos. More generally, any force of nature couples two types of particles to each other, and the larger their charges are the more likely they are to interact (possibly changing into each other in the process). General relativity and quantum field theory seem quite different, yet these two theories meet along a very interesting and puzzling interface: the concept of mass.

It is common to represent the motion of particles on a spacetime diagram, with the horizontal axis representing position, and the vertical axis representing time (past at the bottom, future at the top). A stationary particle traces out a line that runs straight up the page. A quantum of light, or a photon, always moving at the speed of light, traces a diagonal line at a forty-five degree angle across the diagram. In a classical theory like general relativity (or even in special relativity) a particle moving slower than light will trace a line making an angle of less than forty-five degrees with the vertical (time) axis. In quantum field theory, a particle moving, say, from left to right can be thought of as following many different paths all added together. One way to imagine this is to think of the particle jittering back-and-forth, at the speed of light, along its path, sometimes to the left, sometimes to the right. It is, in some sense, the sum of a right-moving particle and a left-moving particle. If you were to trace out its path on a space-time diagram it would zigzag left and right as it moved up the space-time diagram from the past to the future – see (Fig. **3**) (This zigzag picture is spelled out in more detail in Roger Penrose's book "The Road to Reality").

Particles also have a property called *spin.* This name is an unfortunate choice, but for historical reasons we're stuck with it. Rather than thinking of spin as the rotation you associate with a spinning top or a Frisbee, think of it as an arrow (vector) pointing along the axis around which the rotation takes place. The spin axis is defined to be along the arrow, even though the particle itself isn't "really" spinning. And it points in a given direction, just like the Earth's North Pole always points towards the North Star as the Earth orbits the Sun. When a particle is moving in the direction of its spin vector, we call it a right-handed particle. When it's moving in the opposite direction we call it a left-handed particle – see (Fig. **4**).

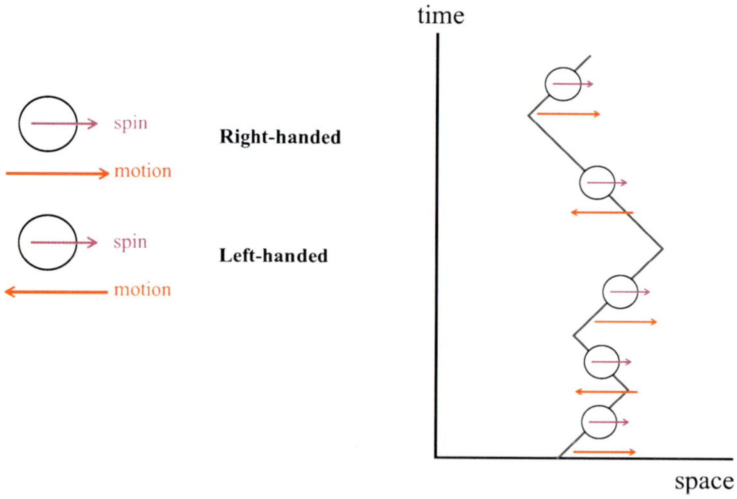

Figure 4: A particle that zig-zags left and right, while its spin remains oriented in one direction will turn from left-handed to right-handed.

Now, returning to that zigzag business. If a particle changes from left-moving to right-moving (or vice versa) while its spin keeps pointing in a fixed direction, it will change from being a left-handed particle to being a right-handed particle (or vice versa). The propensity a particle has for switching between left-handed and right-handed is, essentially, the *mass* of that particle. So, in this sense, mass is the "charge" that couples a left-handed particle to a right-handed particle. A photon, being massless, always moves to the right and travels from left to right as fast as possible. A particle like an electron switches back and forth, sometimes approaching the right-hand-side of the spacetime diagram, sometimes retreating from it, and hence must get from point *A* to point *B* slower than a photon. In quantum field theory, the commonly accepted explanation for the origin of mass is a particle called the Higgs boson. The universe is believed to be pervaded by a field of Higgs bosons, which interact with matter, and each interaction corresponds with a little flip between zigging and zagging, between left-handed and right-handed states. To put it another way, just as an electron can absorb a W^+ particle (or emit a W^-) and turn into a neutrino, a left-handed particle can interact with a Higgs boson and turn into a right-handed particle (or vice versa) – (Fig. **5**).

Many attempts to combine quantum field theory with general relativity postulate another type of boson, the *graviton*. Just as an electrically charged particle is surrounded by a cloud of "messenger" photons, which interact with other electrically charged particles in the vicinity, leading to attraction or repulsion, so too are massive particles surrounded by a cloud of messenger gravitons, which they transmit back and forth between each other, leading to their mutual gravitational attraction.

While the idea of the Higgs boson is part of a theoretical framework which makes extremely accurate predictions about the masses of particles (it was used to correctly predict the masses of the $W^{+,}$ W^- and Z^0 bosons) the Higgs boson itself has never conclusively been observed. Detecting the Higgs boson at an experimental facility like the Large Hadron Collider would be a great triumph for theoretical physics and its predictive power, but at the moment the Higgs boson is just a hypothetical construct. In fact, it may not exist at all (note that even though I refer to "the" Higgs boson, in fact what I mean is a *type* of particle, and not that there is only one Higgs boson in the whole universe). Likewise the graviton may be nothing more than a hypothesis that allows us to calculate in a way that we are comfortable with, but may actually have nothing to do with reality. Personally, I think that proving the existence of the Higgs boson would be a bit of a disappointment. Interesting, yes, but nowhere near as interesting as showing that it most probably doesn't exist. The former result would be a pleasing confirmation of our theories. The latter would force us to recognize that we were on the wrong track, and have to rethink something fundamental. Experimental results that defy our predictions are the School of Hard Knocks for theories.

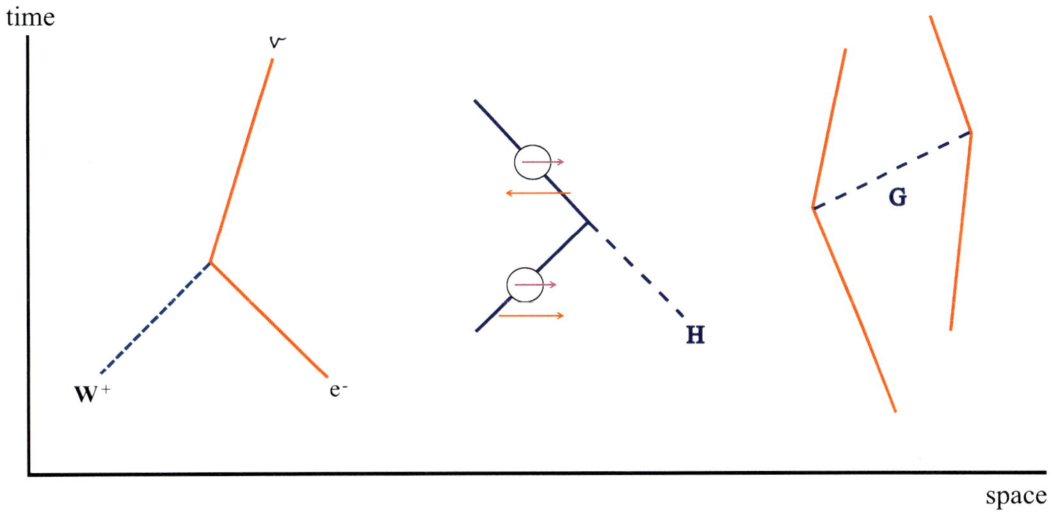

Figure 5: A particle such as an electron (e^-) can interact with a boson (here a W^+) to turn into a neutrino (v) as described in the text. Interacting with a Higgs boson (H) is presumed to give a particle mass, by swapping it between left-handed and right-handed states (center), while two particles may be gravitationally attracted to each other by exchanging a graviton (G) as shown on the right.

All of the above discussion is a rather long-winded way of getting to what I consider to be a very interesting point, namely that the concept of mass which in quantum field theory determines how much an electron, or a proton, or any particle jiggles back and forth between left-handed and right-handed states is exactly the same quantity that determines, in general relativity, how much that particle curves the shape of spacetime. So what does that tell us about combining quantum field theory with general relativity, if we take the facts at face value (like Kepler and Einstein before us)? Perhaps it tells us that the Higgs boson and the graviton are actually not different types of particles, but are really different aspects of the *same* particle. Or maybe that neither the Higgs nor the graviton really exist. And maybe it tells us that the flip-flopping back and forth between left-handed and right-handed states that we refer to as mass is caused by the structure of spacetime itself, rather than by a field of bosons that lives "on top of" some sort of spacetime background. If this is the case, at least some particles (maybe all) are actually features of spacetime, rather than actors moving on the spacetime stage, and a working theory of quantum gravity will actually involve unifying not just gravity with quantum field theory, but also unifying matter with spacetime.

While my personal interests lie along the interface between quantum gravity and particle physics, there are several other directions of enquiry and clues to consider. Perhaps as a colleague of mine, Achim Kempf, of the University of Waterloo and the Perimeter Institute in Ontario, suggests, the success of general relativity as a classical theory is telling us, bluntly, that spacetime can be *both* discrete and continuous at the same time – in much the same way that an electron can be both a wave and a particle. He likens the structure of spacetime to the structure of information on a CD, which is stored in a digital (discrete) form, but which can be used to reconstruct a continuous sound wave. Other researchers have suggested that general relativity is actually in need of modification. It is now quite well established that the Pioneer 10 and 11 space probes, which are on escape trajectories from the solar system into interstellar space, are slowing down faster than expected. It is as though gravity is pulling them back towards the Sun more strongly than calculated. We also know that the stars in other galaxies (as well as our own) are orbiting the centers of their galaxies as if gravity is pulling on them more strongly than expected. These observations suggest the existence of a kind of *dark matter* that exists in haloes around galaxies and increases their total mass, making their gravitational pull stronger. But it may also suggest that our understanding of gravity simply needs to be modified. If Einstein 's ideas need to be corrected when applied to gravity acting on interstellar scales, then a successful theory of quantum gravity will need to describe these modifications, or else it will be an incomplete description of reality.

Furthermore, despite the fact that gravity seems to be stronger than expected, the universe in its entirety seems to be expanding at an ever-increasing rate, like a self-inflating balloon. It almost seems as if spacetime didn't come into existence only at the Big Bang, but is always being spawned anew from the spacetime that already exists, in an exponential process. If this is the case, the behavior of the universe at the largest scales is telling us something about the behavior of spacetime at the smallest scales. But – and here's the large pinch of salt – that's just my hunch. Every anomaly, every unexpected observation (if it can be properly confirmed by repeated measurements) is a clue that tells us something about the universe we inhabit. To paraphrase Carl Sagan : the universe is often obscure but it never lies. We are faced with the frustrating and exciting task of figuring out what the universe is telling us, but we can't do it without a proper balance of theory to speculate on the nature of reality, and experiment to ensure that reality is indeed what we're speculating about. It is precisely this that makes science unique, powerful, and interesting.

CHAPTER 8

Shooting in the Dark

Sabine Hossenfelder

Nordita, Roslagstullsbacken 23, 106 91 Stockholm, Sweden

Abstract: Why are we looking for a Theory of Everything, what do we mean by that, and how do we make headway on it? After addressing the why and what, we briefly survey the phenomenological models used and the search for experimental guidance.

"The more important fundamental laws and facts of physical science have all been discovered, and these are so firmly established that the possibility of their ever being supplanted in consequence of new discoveries is exceedingly remote"

~ Albert Abraham Michelson, 1903 [1]

INTRODUCTION

Only a few years after Michelson wrote the above, Einstein 's theory of general relativity and the theory of quantum mechanics overthrew our understanding of nature. The following century brought stunning progress in physics and with it technological advances that consolidated fundamental research as a driver of innovation.

This success rests heavily on us today. The better physicists explained observations, the more remote the possibilities became for new discoveries. A century ago, table-top experiments and studies designed and completed by small groups in their own facilities were sufficient. Michelson 's interferometer, the measurement of the photoelectric effect, or the invention of the laser are examples that made history.

But the unknown has moved away from us, towards larger telescopes, more powerful particle colliders, billion-dollar satellites, and cosmic ray telescope arrays covering hundreds of square kilometers all with the purpose of revealing nature's next secret. Today, physicists routinely form collaborations with thousands of members, and new experiments require tremendous amounts of planning and coordination, not to mention money.

It is thus not too surprising that progress has slowed down. But it is even more infuriating since, unlike Michelson, we know that the fundamental laws of physics we have discovered are incomplete.

WHY

At the beginning of the 21st century, we have an extremely successful set of theories: the Standard Model of particle physics and general relativity. With these two theories we can explain observations and predict experiments to an extremely high degree of precision. But in contrast to the Standard Model which combines the strong, weak, and electromagnetic forces in the framework of quantum field theories, general relativity is an entirely classical (non-quantum) theory. This axiomatic difference is not only unsatisfactory for the esthetically inclined physicist who strives to pursue the path of unification, but it is also much worse in that we cannot use both theories together the way they are, while we know that nature does exactly that.

Quantum particles can exist in superposition states, i.e. in different, sometimes contradictory states, at the same time. The best known example is Schrödinger 's cat which is neither dead nor alive. A more realistic example is a photon traversing a double-slit that is neither here nor there. These animals, or particles respectively, carry energy and thus gravitate. Yet we do not know what their gravitational field is: as a classical field it does not exist in superposition.

Thus arises the need to quantize gravity. Gravity however has resisted all such efforts for more than half a century. Though, to be precise, the problem is not in quantizing gravity per se, but rather in that doing so using the same methods applied to the other interactions yields nonsensical results.

Moataz H. Emam (Ed)

There are two additional arguments commonly mentioned for why a quantization of gravity is necessary. One is that general relativity predicts the formation of singularities; instances of infinite energy density and gravitational forces. Such singularities are usually considered unphysical and signal a breakdown of the theory. In these extreme regions of spacetime, general relativity would have to be replaced with a more fundamental theory.

The second, related, argument is the so-called black hole information loss problem. Using quantum field theory in a classical black hole geometry, Hawking [2] showed that black holes emit thermal radiation and thereby lose mass. If this radiation remained thermal until the black hole had entirely evaporated, then any matter distribution with the same mass that collapsed to a black hole would eventually be converted into the same thermal final state. Detailed information contained in the initial configuration would have gotten lost. Such an irreversible loss of information is incompatible with quantum mechanics. One hopes that a proper quantization of gravity will solve this contradiction[1].

These two latter arguments are based on weaknesses in our current understanding that could be solved by a theory of quantum gravity, while the previously mentioned problem of superposition states is a stronger argument in that its solution necessitates the existence of a quantum theory of gravity.

In addition to these theoretical reasons to look for quantum gravity, we also have experimental data which is insufficiently understood, and thus points at shortcomings in our current theories. We know that the biggest part of what fills our universe is not the usual stuff that we see around us every day. Instead, an increasing amount of cosmological and astrophysical observations show that 95% of the universe's content has peculiar properties, different from those of the matter we are made of. These new types of stuff have been dubbed *dark energy* and *dark matter*. We have models explaining the observations, but we do not know what the microscopic origin of the dark stuff is, or whether it is some sort of "stuff" at all. It has repeatedly been tried to find the origins of dark energy, for example, in quantum gravitational effects. Other types of unexplained data include, rather banally, the masses of elementary particles. Why they have the values we experimentally observe nobody knows. Likewise nobody knows why the subatomic particles can be classified in the three categories known as generations. It is far from clear a theory of quantum gravity would explain these puzzles, but it could give us a clue on how to proceed.

WHAT

Let us be clear about what we mean by the expression "quantum gravity". We use it for any framework able to resolve the apparent disagreements between quantum theories and general relativity. This does not necessarily imply a quantization of gravity, though it is the most widely pursued path. A different point of view that has become more common during the last decade is that the gravitational interaction might instead be merely a macroscopic theory and that quantizing spacetime will not result in a fundamental theory any more than quantizing water waves would result in a theory of atomic structure [4, 5]. Such approaches, collectively known as *emergent gravity*, we will subsume in the phrase "quantum gravity ".

There also exists the possibility that the theory of quantum gravity would give rise to the Standard Model of particle physics, and thus explain, among other things, the masses of elementary particles. This is a goal vastly more ambitious than just trying to marry quantum field theory with general relativity, but as an additional requirement also provides guidance. A theory that achieves a consistent quantization of gravity and entails the Standard Model of particle physics is commonly referred to as the (yet-to-be-found) *Theory of Everything*.

Though widespread, the expression "Theory of Everything" is an unfortunate misnomer. Such a theory may describe "everything" we know today (and even that is not certain), but we could never be sure it would eternally continue to describe all observations, thus we would never know it was indeed truly a theory of everything. Theories are scientific exactly by virtue of being falsifiable through new evidence, thus claiming a theory to be a theory for everything is unscientific in itself.

---

[1] This hope is not without problems either, but further discussion would lead us astray. One can also argue that the problem with singularities and that of black hole information loss are actually the same problem, see [3].

It would also most certainly *not* be a theory that actually describes all natural phenomena. Already in physics there are many levels of emergent phenomena that are not derivable from a fundamental theory, not in practice and possibly not even in theory [6]. And it gets only more involved in highly complex systems. From physics to chemistry, biology, psychology and sociology, a true "Theory of Everything" would have to explain, well, everything, and it is doubtful unifying the Standard Model and general relativity would achieve that.

However, examples of misleading expressions in common use are legion; take any region or city with the word "new" in its name that isn't new any more than a theory once thought to explain everything might still do so a century later. Complaining about it is thus moot and we will continue to use the expression "Theory of Everything" with the understanding that it means a unification of all of the presently known interactions in a single framework, necessarily including some version of quantum gravity. It remains to hope that future generations will choose names more modestly.

HOW

Now that we have reviewed what physicists are searching for and why, let us see how they do so. A big obstacle on the way to a Theory of Everything is lacking experimental accessibility since the gravitational interaction and thus its quantum effects are very weak. Gravity might not appear such a weak force if it keeps you in your chair, but the only reason we notice it so prominently in our daily lives is that, unlike the other interactions, it cannot be neutralized. The electromagnetic interaction of something as small as a fridge magnet is already stronger than the gravitational pull of the whole Earth and in addition even holds your pizza flyer.

Extrapolating the strength of gravity to very small distances, one finds that its quantum effects would become relevant at a scale of 10^{-33} cm, which corresponds via the Heisenberg Uncertainty Principle to an energy of 10^{16} TeV. Known as the Planck scale, these values are far beyond what we have been able to test in Earth-based experiments, and what we will likely be able to test in the foreseeable future. For comparison, the Large Hadron Collider (LHC), a gigantic project that started operating at the end of 2009, will reach 14 TeV, which corresponds to 10^{-18} cm, 15 orders of magnitude below the Planck scale. A collider to test quantum gravity would need to be the size of our galaxy, and even a detector the size of the planet Jupiter wouldn't detect as much as a single graviton [2] in the lifetime of the universe [7]. This led Freeman Dyson to hypothesize that the regime in which the Standard Model and general relativity lead to conflicting results is entirely undetectable [8].

This depressing situation has put the ball into the field of theoretical physics. There seem to be two ways to venture forward into the unknown from where we presently are. One way is a top-down approach; the attempt to develop a new theory from first principles. The problem with this approach is that not only does it typically imply the development of entirely new mathematical techniques, but the new theory must also be successfully connected to the theories and observations we already have. Both major top-down approaches, string theory and loop quantum gravity, are currently stuck in that phase. Top-down approaches put a strong emphasis on mathematical consistency since it is, besides inspiration, the only guide at hand.

The other way is a bottom-up approach, starting from the theories we already have and trying to extend them with the hope of discovering a path leading forward. For example, one may allow for violations of Lorentz-invariance – the symmetry of Einstein 's theory of special relativity – and then constrain all subsequent effects, or one may attempt modifications of general relativity by adding higher order curvature contributions. The problem with this approach is that without any guidance there are many different ways to go. Bottom-up approaches are typically considered to be mere bridges towards a more fundamental theory and thus the emphasis on mathematical consistency is not quite as strong as with the top-down approach. One can expect them to leave unanswered questions for the fundamental theory to explain.

Taking clues from top-down approaches and using them as particular bottom-up extensions is then a promising middle-path to take, and we will in the following discuss some examples that one could call "top-down inspired

[2] A graviton shouldn't be confused with a gravitational wave. The former is a consequence of a quantized theory of gravity, while the latter is an entirely classical phenomenon. None of which should be confused with *gravity waves*.

bottom-up approaches." While not meant to be fundamental, the so constructed models aim to predict a phenomenology that tests for a specific property the fundamental theory might have. These efforts have been lately dubbed "phenomenology of quantum gravity".

PHENOMENOLOGICAL QUANTUM GRAVITY

The most obvious place to look for effects of quantum gravity is the early universe, where the gravitational interaction must have been very strong. There, quantum gravity could have left traces in the cosmic microwave background. And while these traces are still outside today's measurement precision, it might not be that much longer for the data to become available.

Quantum gravitational effects are also very strong near the singular center of a black hole, where collapsing matter is crushed to extremely high densities. These regions of spacetime were thought hidden from observation by the presence of an event horizon, preventing causal contact with the interior. However, recent analysis indicates that collapse without the censorship of a horizon is far more common than originally thought [9]. The quantum gravitational region could thus remain in causal contact with the observer. It has also been shown [10] that, under certain circumstances, a collapse does not lead to a singularity at all, but to a dense object without a horizon that nevertheless displays strong quantum gravitational effects at its surface. Exactly how one can extract useful knowledge from matter passing through or originating from these regions is still under investigation, but these are interesting developments telling us that quantum gravitational effects are not as inaccessible as we had originally thought.

Another very generic expectation from a theory of quantum gravity is fluctuations of the background geometry that could affect the propagation of particles, possibly inducing violations of symmetries of the Standard Model (known as T, CP, CPT) or stochastic deviations from Lorentz-invariance. These effects might be observable even in Earth-based high-precision experiments.

Background fluctuations would also make themselves noticeable as noise in gravitational wave interferometers [11]. With additional modifications to the model leading to such fluctuations, such as [12], the noise could just be within the precision of the currently operating GEO600 experiment [13]. This also exemplifies a common tension in the phenomenology of quantum gravity: The conservative and straightforward approaches are very hard to test. Additional assumptions can be used to shift effects into the (almost) accessible range, but these assumptions also make the model more speculative and less plausible.

Departures from Lorentz-invariance are indeed a promising signature of quantum gravity and one that is presently particularly exciting because the precision of experiments under way already allows constraining known models. Deviations from Lorentz-invariance can lead to the propagation of photons being affected depending on the photon's energy, such that the speed of light changes with the energy of the photon. This effect is tiny even for photon energies in the GeV[3] range, but it does add up over the distance traveled. The ideal photon source is thus far away and highly energetic, which makes gamma ray bursts excellent candidates to observe such effects.

In fact, the Fermi telescope, which was launched in June 2008, has detected a much larger number of very high energy gamma ray bursts than previously expected. And the arrival of photons from these bursts also offered surprises. Several bursts have now been documented in which the higher energy photons (>GeV) arrive with a delay of more than 10 seconds after the onset of the burst has been observed in the low energy range (keV-MeV). While it is still unclear whether this delay is caused at emission or during propagation, more statistics and better analysis – in particular of the delay's dependence on the distance to the source – will eventually narrow down the possible causes and constrain models that give rise to such features [14].

A completely different category of models studies the possibility of quantum gravitational effects being much stronger than usually thought due to a modification of the gravitational interaction on very short distances. Such a modification occurs in some scenarios with extra spatial dimensions, whose existence is predicted by string theory, and has the consequence that quantum gravity could become observable in Earth-based collider experiments, such as the LHC.

---

[3] *Editor's note*: Gigaelectronvolts. See footnote 15 on page 34

If this should turn out to be a correct description of nature, we would see the production of gravitons and black holes in the LHC [15]. The gravitons themselves would not be captured in the detectors and would lead to a missing energy signal. Black holes would decay via Hawking radiation. Ideally the distribution of decay products would allow us to determine the parameters of the model and the number and size of the extra dimensions. Black hole production and decay would be a striking signature, and allow us to examine in the laboratory the fate of black hole information during the evaporation process.

It should be emphasized that the above scenarios make use of phenomenological models that are, at least so far, not derived from any of the presently pursued approaches towards quantum gravity. The purpose of these models is to study consequences that arise from specific features the underlying theory could have and constrain them. In such a way, we can learn about the general properties of the theory we are trying to find, for example whether it does have additional spatial dimensions, realizes a holographic principle, or results in deviation from Lorentz-invariance.

SHOOTING IN THE DARK

The Phenomenology of quantum gravity is a very young research field exploring a possibility which unfortunately has been neglected for too long: the extraction of information about the unknown, sought- after theory from experiment. All of these efforts are shots in the dark, guesses based on little or no information, and it is likely many of them will miss the target. The ideas put forward and models used are in their infancy and many are in need of improvement. But it is good to see this field becoming integrated into the so-far mathematically dominated search for a Theory of Everything.

REFERENCES

[1] A. Michelson, *Light waves and their uses*. University of Chicago Press, 1903.

[2] S. Hawking, "Particle creation by black holes", Commun. Math Phys., pp43:199-220, 1975.

[3] S. Hossenfelder and L. Smolin, "Conservative solutions to the black hole information problem", *http://arxiv.org/abs/0901.3156.*

[4] T. Jacobson, "Thermodynamics of spacetime", *Phys. Rev. Lett.,* 75, pp1260-1263, 1995.

[5] T. Padmanabhan "A dialogue on the nature of gravity" *http://arxiv.org/abs/0910.0839.*

[6] M. Gu, C. Weedbrook, A. Perales and M. Nielsen, "More really is different", *Physica* D238, pp835-839, 2009.

[7] T. Rothman and S. Boughn, "Can gravitons be detected?" *Found. Phys.,* 36, p1801, 2006.

[8] F. Dyson, *The Scientist as Rebel*. New York Review Books, 2006.

[9] P. Joshi, "Do naked singularities break the rules of physics?" Sci. Am., January 2009.

[10] C. Barcelo, S. Liberati, S. Sonego and M. Visser, "Revisiting the semiclassical gravity scenario for gravitational collapse", *http://arxiv.org/abs/0909.4157.*

[11] G. Amelino-Camelia, An interferometric gravitational wave detector as a quantum gravity apparatus", *Nature* 398: pp216-218, 1999.

[12] C. Hogan, "Inteterminacy of holographic quantum geometry", *Phys. Rev.* D78, p087501 2008.

[13] Geo600 website: *http://geo600.aei.mpg.de/*

[14] G. Amelino-Camelia and L. Smolin, "Prospects of constraining quantum gravity dispersion with near term observations", *Phys. Rev.* D80: p084017, 2009.

[15] G. Landsberg, "Collider searches for extra spatial dimensions and black holes", *http://arxiv.org/abs/0808.1867.*

CHAPTER 9

Deciphering Quantum Theory

P. Goyal

Perimeter Institute for Theoretical Physics, Waterloo, Ontario, Canada; E-mails: pgoyal@perimeterinstitute.ca; philip.goyal@gmail.com; http://www.philipgoyal.org

Abstract: Quantum theory poses deep challenges to the mechanical conception of reality that underlies classical physics. Yet today, over eighty years after its creation, its implications for our picture of reality remain enshrouded in uncertainty. In view of the current search for a more comprehensive theory of physics (a so-called theory of everything), it is vital that these implications be clearly elucidated. In this article, I describe the nature of the challenge posed by quantum theory, and outline efforts that have been made to better understand its non-classical features, such as non-locality. In particular, I discuss the informational perspective, which, through the study of quantum information processing, has provided deep insights into the nature of quantum reality, and has also revitalized the long-standing quest to reconstruct the content of the rather mysterious mathematical formalism of quantum theory from a set of crisp physical principles. Finally, I indicate some implications of recent reconstructive work for the search for a theory of quantum gravity, and, more broadly, for our picture of physical reality.

INTRODUCTION

Quantum theory is perhaps the most empirically successful theory in the history of physics. In the eighty-five years since its creation, it has proven itself capable of accounting, to a high degree of precision, for a vast range of physical phenomena such as the principles of chemical bonding and chemical reactions, the nuclear reactions that fuel the stars, and exotic phenomena like superconductivity. Much of the modern technology that fuels our lives is underpinned by quantum theory. The transistor; the basis for the modern computer that underlies all information processing technology, requires quantum theory for its design and modeling. The same holds true for the laser and the light-emitting diode, which jointly provide the basis for optical communication networks (the backbone of the telecommunications industry) and optical data storage (in the form of CDs and DVDs). Yet more quantum technology, such as quantum computers and quantum cryptography, is waiting in the wings, promising to transform our lives still further.

However, ever since its very beginnings, quantum theory has been of great interest for quite another reason. Taken at face value, quantum theory challenges many of the key tenets of the mechanical conception of reality. That picture of reality, which was developed by Descartes, Galileo, and Newton (amongst others), underpinned the development of physics—what is now called *classical physics* —for approximately three hundred years. So successful was this enterprise that it spawned the so-called *mechanical world view*, which had a profound effect not only on other scholarly disciplines, but also on the general culture. Thus, quantum theory forces reconsideration of not only a view of reality of interest to physicists, but also of a world-view that has guided the development of Western society for more than two centuries.

For physicists, the challenge posed by quantum theory is, at a minimum, to develop an intuition for the reality that it describes which is sufficient to be able to discover, explore, and harness the phenomena it encompasses, and, at best, to develop a conception of physical reality which takes quantum theory fully into account, to develop a conception which is as coherent and compelling as the mechanical conception of reality, and which is capable of guiding further development of physics, particularly the current search for a so-called theory of everything. For the philosopher and thinker, the broader challenge is to develop a conceptually coherent and plausible world-view that fully takes the content of quantum theory into account. In this article, I shall attempt to sketch the degree to which the challenge to understand quantum theory has been met and the means by which this has been done, and to give some sense of the road that lies ahead.

CLASSICAL PHYSICS

To understand the challenge posed by quantum theory, it is helpful to begin by considering classical physics. Underpinning classical physics is a mechanical conception of reality. This conception is multi-faceted, but its essential "attitude" is that all particular events that occur in the physical world are, in their finest details, manifestations of general principles of nature. The promise is that, by careful study of particular events, we, in spite of our limitations as finite beings, can discover these general principles by a process of generalization. More precisely, the mechanical conception posits that the totality of all that exists in the phenomenal world (namely that aspect of the world registered by our senses or instrumental extensions thereof such as microscopes or telescopes) is *matter* moving on the *fixed stage of space* in step with a *universal time*. That is, all that exists in the physical world is matter in motion, the manner in which it moves being completely governed in quantitative detail by *universal laws of motion*. Thus these laws, as it were, stand outside the Universe dictating all that happens within it, but are left unaffected by what happens. It is further posited that, in principle, observers can probe this matter as precisely as they wish without disturbing its nature or its motion. Thus, in principle, observers have a "God's eye" view of reality: everything, in its finest details, is open to their gaze. Moreover, reality is so constructed that, even though observers are necessarily restricted to studying *parts* of the universe, they can nonetheless in principle come to learn the nature of the universal laws of motion, and thereby aspire to a theoretical model of physical reality that allows arbitrarily precise prediction of the behavior of the entire physical universe.

In the hands of Descartes, Galileo, and Newton, this conception was mathematically formulated in a fairly direct way. In Newton's mechanics, the first comprehensive theory of classical physics, published in 1683, matter is composed of particles represented by geometrical points in Euclidean space (i.e. a space that follows Euclid's geometry), moving in geometrically precise trajectories determined by a set of equations of motion in step with a universal time. The dynamics are such that all information about any time in the past is contained in the present. This theory was enormously successful, accounting in a unified way for the known regularities of the solar system (codified as Kepler 's laws) and the precise trajectories of projectiles (such as cannon balls) on the surface of the Earth, and also making new predictions (such as the arrival of Halley 's Comet in 1758). Over the subsequent two hundred years, theories of electricity and magnetism (Electromagnetism), and of heat and temperature (Thermodynamics), were developed and were similarly successful. These theories stretched the mathematical framework established by Newton. For example, whereas Newton believed that all that exists is particles in motion, which means that it requires only a finite (or countably infinite) number of parameters to fix the state of these particles, Faraday and Maxwell introduced the idea that there exist dynamical *fields* which require *continuously* many parameters for their specification. However, these modifications did not fundamentally challenge the mechanical *conception* of reality of classical physics [1].

It is important to note that this "clockwork universe" view of the phenomenal world was also readily graspable by those who had no knowledge of mathematics. Consequently, this view was not confined to the physics community, but propagated to other areas of science (such as chemistry, biology, psychology, economics, and sociology) which were able to adapt this view to their own domains of enquiry and were also often able to successfully adapt the mathematical techniques developed by physicists. More broadly, this view also propagated beyond the sciences to areas such as philosophy and theology, where it had a profound impact which, in the more than 300 years since Newton announced his mechanics, has played a vital role in the radical transformation that has occurred in our understanding of our place in the world.

QUANTUM PHYSICS: BEGINNINGS AND EARLY IMPLICATIONS

By the 1890s, classical physics was in a highly developed state. However, there were some clouds on the horizon – curious experimental facts which stubbornly resisted explanation within the framework of the existing theories of classical physics. One of these facts concerned the frequencies of light given off by heated bodies. The classical theories were unable to account even *qualitatively* for the experimental data. The tentative explanations of these

[1] Perhaps the singular exception was the threat posed by the second law of thermodynamics, which asserted that certain physical processes are *irreversible*, which conflicted with Newton 's assertion in the fundamental reversibility of all dynamical processes. This remains a contentious point today, but has been overshadowed by the challenges posed by quantum theory.

stubborn facts, initiated by Max Planck, would, over the next 30 years, bring about the development of quantum theory – an entirely new theoretical foundation for physical theories that would replace the mathematical framework of classical physics.

Quantum physics did not come into existence in a straightforward and principled manner. Unlike the mathematical framework of classical physics, it was not arrived at by mathematically formulating some clear conception of reality. Rather, it arose through a somewhat tortuous path consisting of rather *ad hoc* modifications of classical physics, guided by heuristic ideas (an example of such idea, due to de Broglie, is that each particle has an associated "guiding wave") and by ingenious mathematical guesswork about what aspects of the existing mathematical structures of classical physics needed to be changed and what aspects could be retained.

The first quantum theory was non-relativistic quantum mechanics, developed in distinct forms by Schrödinger and Heisenberg in 1925–6. Its key achievement was accounting in quantitative detail for the light spectrum of hydrogen in a variety of different circumstances. Over the course of the subsequent few years, a general and rather beautiful mathematical formalism – the *quantum formalism* – emerged, which provided a new mathematical framework (replacing the mathematical framework of classical physics) within which to build physical theories.

Owing to the indirect process by which the quantum formalism was created, the physical origin of many of the mathematical aspects of the formalism was not clear. Nevertheless, some physical features could safely be read off the formalism, and these were sufficient to establish (within a few short years after its creation) that the theory departed radically from the framework of classical physics.

As mentioned above, in the classical framework an observer is afforded a God's eye view of physical reality. Thus, an observer can perfectly and completely access the goings on (or the *state)* of a physical system, and furthermore can do so without disturbing the system to any significant degree. In contrast, the quantum formalism posits a new model of the measurement process which has three distinct features:

1. *Probabilistic Outcomes.* The outcomes of measurements performed on systems are only predictable on a probabilistic level.
2. *Complementarity.* A single type of repeatable measurement only yields information about some aspect of a system, at the expense of some other aspect of it.
3. *Discreteness.* Measurement outcomes can be discrete, being finite in number or countably infinite.

The first of these features is that, in contradistinction to the fundamental attitude of the mechanical view of reality, it is *not* true that every detail of every event is determined by universal laws. Instead, the quantum formalism asserts that only the *probability* that a measurement will yield a particular result is predictable; the outcome that will be obtained in a particular run of an experiment, in general, is not. The quantum formalism, as does classical physics, considers measurements which are repeatable – that is, measurements which, when immediately repeated, yield the same outcome with certainty. If one considers such measurements, then it follows as a direct consequence of the fact that measurements are probabilistic that, in general, they *disturb* the state of the system upon which they are performed. In fact, they disturb it almost completely – almost no trace of the pre-measurement state of the system is left in the post-measurement state.

The second feature, complementarity, can be precisely expressed in a number of different ways. Perhaps the simplest is to say that, unlike the situation in classical physics, one cannot perform a repeatable measurement on a system which yields information about *all* of the degrees of freedom of the state of the system[2]. Together, these two features severely constrain an observer who wishes to learn about the unknown state of a physical system. First, due to complementarity, an observer cannot rely upon one type of measurement, but must use more than one. Second, due to probabilistic outcomes, an observer must perform many measurements on identical copies of the system. Yet, after a finite number of measurements, the observer's knowledge about the state will still be imperfect. It is only in the unattainable, idealized limit of an infinite number of measurements that the observer's knowledge of the state

[2] In fact, if a system is in a so-called *pure state*, a repeatable measurement only accesses *one half* of the degrees of freedom of the state.

becomes complete. Therefore, there is almost always an *informational gap* between the theoretical description of the underlying reality – the quantum state of the system – and the observer's knowledge of that state.

The third feature of the measurement process, discreteness, is that, when certain measurements are performed on physical systems, the number of possible outcomes can be finite (or countably infinite) in number[3]. This stands in contrast with the classical assumption that all physical quantities (such as the position of a particle) can take a continuum of possible values. Hence, discreteness challenges the classical idea that the continua of space and time are the fundamental bedrock of physical reality.

One of the basic premises of classical physics is that one can conceptually *decompose* the whole of physical reality into spatially disjoint *parts,* describe each of these parts separately, and then combine these partial descriptions together to form a description of the whole. This premise is required in order that an observer, who is necessarily restricted to studying the physical universe one piece at a time, can nonetheless aspire to a complete description of reality. However, the quantum formalism asserts that physical reality is *not* constituted in this way. For example, consider the simplest case of a physical system consisting of two spatially separate subsystems. The quantum formalism asserts that almost all possible states of the system *cannot* be specified by giving the state of each subsystem separately. Such a non-decomposable state is technically referred to as an *entangled* state, a term coined by Schrödinger. More generally, quantum theory implies that a physical system can *become* entangled with other systems as it interacts with it. Consequently, one can expect that a typical physical system as it is found in nature is generically entangled with many other physical systems, including those at great distances from it. However, an observer studying that system has no way of determining the nature of these entanglements by studying the system alone.

Entanglement also has another important ramification. In classical physics, it is assumed that a system can only affect another system through influences that propagate through space at some finite speed (the speed of light), an assumption known as *locality*. However, Schrödinger showed that, according to the quantum formalism, if an entangled state of two subsystems is shared by two parties, Alice and Bob, then Alice can instantaneously affect – or *steer* – the state of Bob's subsystem by performing a measurement on *her* subsystem, an effect whose strength is moreover independent of how far away Bob's subsystem happens to be! Thus, the quantum formalism also appears to undermine the classical ssumption of locality in a fundamental way. However, the situation is not as simple as one might expect: as Schrödinger also showed, Alice cannot use steering to instantaneously *signal* to Bob. Thus, locality is preserved at the level of signaling[4].

THE MEASUREMENT PROBLEM

In classical physics, a measurement performed on a physical system can be regarded as an essentially passive recording of information about the system: in the ideal case, information is gained without in any way affecting the system. In contrast, as described above, according to the quantum formalism the process of measurement is an active intervention in the world: in general, the state of a system will be changed as a result of a measurement being performed on it. One immediate consequence of this difference is that, in the quantum formalism, a physical system can undergo two *kinds* of physical process, mathematically described in quite different ways. The first is deterministic temporal evolution, which is familiar from classical physics, and represents the change in the state of the system as a result of the passage of time while the system is in some "environment" (consisting of other physical systems) with which it is interacting. The second is a nondeterministic (probabilistic) change due to a measurement being performed on it. But now the question arises: what kind of interaction with a physical system should be treated as a deterministic evolution, and what kind of interaction should be treated as a measurement?

The quantum formalism is silent on this question. The formalism itself talks of "measurement" purely in the abstract, leaving the physicist to decide when to describe an interaction as a measurement. In *practice*, this ambiguity in the rules of application of the formalism is rarely a problem: An experimental physicist sets up an experiment, which, if it is to provide information about what is going on, necessarily contains some kind of a measurement device such as a photographic screen or similar detection device, and this device demonstrably

[3] For example, a measurement of the magnetic moment of a silver atom will yield one of just *two* possible outcomes.
[4] Special relativity forbids instantaneous, faster than light, signals.

produces outcomes. However, as a matter of *principle,* this ambiguity is troubling, and has produced sufficient disquiet so as to become known amongst physicists as the *measurement problem.*

QUANTUM THEORY AND OUR VIEW OF PHYSICAL REALITY

In the period shorlty after the formulation of quantum theory, physicists responded to the above non-classical features of quantum theory in essentially two distinct ways. The first view, held by many of the founders of quantum theory, such as Bohr, Heisenberg and Pauli, was that the non-classical features of the formalism, such as the statistical nature of its predictions, had to be taken seriously. That is, these features reflected the very *structure* of physical reality, and the classical mechanical view had to be replaced by something fundamentally new. In order to illuminate these non-classical features of the quantum formalism, some of these physicists attempted to identify related concepts in the existing philosophical literature or to develop new concepts of which these features could be regarded as particular instantiations, thereby placing these non-classical features in a broader philosophical light. For example, Bohr developed the concept of *complementarity*, which he expressed as meaning that the process of coming to know anything about some aspect of reality unavoidably has the effect of bringing into existence some aspect of it at the expense of simultaneously rendering inaccessible some other aspect of it [1]. Bohr believed that this concept had general validity beyond the field of physics itself, in areas such as psychology and sociology for instance [2]. In his view, the impossibility in quantum theory of performing a repeatable measurement that provides complete knowledge about a physical system (but only one property of it at the expense of another) was simply a special case of the general principle of complementarity. Similarly, Heisenberg and Pauli both suggested that the Aristotelian notions of potentiality and actuality could be useful in understanding the relationship between the state of a system and the outcomes of measurements performed upon the system, and in understanding why the classical mode of thought led to inconsistencies when applied to such simple experimental situations as Young 's double slit experiment [3, 4].

The second view, held by Einstein and some other physicists, maintained that notwithstanding the manifest non-classical features of quantum theory, the classical mechanical view of reality did not require the revision of its fundamental tenets. Einstein, for instance, argued that the statistical nature of quantum predictions was simply an indication that the quantum description of reality was incomplete, and spent a significant part of his later life searching for a classical field theory that was capable of underpinning quantum theory. de Broglie, in particular, supported this point of view by showing that the quantum theory of an ensemble of particles could be re-written in a form closely akin to classical mechanics, albeit with some curious non-classical features.

It may seem surprising that such different viewpoints could be taken based on the same evidence. However, it should be borne in mind that, at its core, a physical theory is simply a set of mathematical equations, and rules for their application. A theory does not "tell" us anything about the nature of reality itself. Every statement one makes on the basis of a physical theory about the nature of reality itself is an extrapolation, and is strongly dependent upon the metaphysical inclinations of the individual. Thus, in the face of quantum theory, it was perfectly rational for physicists to adopt quite different viewpoints as to its implications for our view of physical reality.

DECIPEHRING QUANTUM THEORY

Today, some eighty years after the early efforts to understand quantum theory, there is still no broad agreement on what conclusions about the nature of reality we ought to draw from quantum theory. Nonetheless, in the interim, a number of important advances have been made. Two of the most striking developments have been the exploration of quantum non-locality, and the development of the *informational* perspective on physical reality.

As mentioned above, Schrödinger 's work on entanglement showed that, at the level of theoretical description, two subsystems cannot in general be regarded as separate entities. This finding *suggests* that physical reality is non-local in the sense that an experimenter performing a measurement on one subsystem can instantaneously affect the state of the subsystem with which it is entangled, even if the subsystems are widely separated in space. But, it *could* be that the non-separability one sees in the quantum formalism is simply an artifact of the formalism, and not a reflection of physical reality itself. This latter possibility is supported by Schrödinger's demonstration that two entangled subsystems cannot be used to send instantaneous signals. However, at the time of Schrödinger's work, given that it did not appear to be possible to experimentally *test* for the existence of non-locality, it was difficult to choose between these two possibilities. This deadlock was broken in an extraordinary fashion in 1964 by John Bell [5]. Bell showed, that, if a pair of quantum systems is allowed to interact and are then separated by an arbitrarily great distance, the outcomes of measurements performed on each of them separately can be *correlated* in a way that cannot be accounted for if we assume that reality behaves in accord with the mechanical conception of classical physics (being deterministic and local, in particular). In short, Bell showed that entanglement does indeed leave an experimentally detectable fingerprint which cannot be accounted for based on a so-called "local realistic" picture of reality.

Bell's theorem strongly suggests that there is *some kind of connection* between physical systems that have interacted in the past, a connection that endures irrespective of their distance and which enforces subtle but very real correlations between them. This connection is of a type never seen in classical physics. Unlike the forces of gravity, electricity or magnetism, the connection is un-attenuated by distance, and is specific to the systems which previously interacted. Nevertheless, Bell's theorem rests on a number of assumptions, and one can rationally question the validity of some of these assumptions, and thereby avoid the conclusion that there is some kind of non-locality in nature.

In recent years, as we shall describe below, the conclusion of Bell's work has been bolstered by the demonstration that entanglement is a powerful physical resource that can be used to carry out various processes that are impossible or very difficult classically. If true, non-locality fundamentally alters the status of *space* and with it a key element of the mechanical view; space is no longer the entity which mediates all interactions between material bodies; instead, bodies also interact via an ever-changing web of inherently non-spatial connections.

INFORMATION PHYSICS

The concept of information is an intrinsic part of our everyday experience. Each of us is an observer immersed in the physical world, and our knowledge of that world (including the knowledge possessed by other observers) is necessarily limited. Consequently, we must constantly ask questions of the world and use our sensory systems in order to gather more information about it, and make inductive inferences about the world based on limited information. Yet, despite its ubiquity in everyday experience, the concept of information is conspicuously absent in the theories of classical physics (such as classical mechanics and electromagnetism). The reason for this is two-fold. First, in the classical framework, the universe evolves *deterministically* and *reversibly* – the past fully determines the future and, conversely, the future determines the past, so that no information is lost or gained over the passage of time. Second, observers can always, in principle, perform measurements that yield perfect and complete information about the state of reality. Therefore, every observer in principle has the same "God's eye" view of reality. If an observer lacks knowledge about the state of reality, then that is the result of contingent circumstances (such as due to limited access to experimental devices) rather than reflecting a fundamental feature of reality and the observer's relation to it. For these two reasons, the concept of information is essentially redundant in classical physics.

However, as mentioned earlier, the quantum formalism posits a new model of the measurement process, from which it follows that there is almost always an informational gap between the observer's knowledge of the state of a physical system and the state itself. Furthermore, when an observer performs a measurement to obtain information about a system, the state of the system is, in general, irrevocably altered. Hence, there is, in general, no information gain without disturbance. Thus, in conceptualization of the quantum formalism, the concept of information naturally occupies a central role.

Over the last few decades, these non-classical features of quantum theory, together with developments in theories of inference (information theory and probability theory) and the rise of information technology in everyday life and the concomitant rise of computer science as a discipline, have given rise to the belief amongst many physicists that the concept of information may have a critical role to play in our understanding of quantum theory. More broadly, in recent years there has been considerable interest in the program, perhaps best articulated by John A. Wheeler under the slogan *"It from Bit"* [6, 7], of providing an information-theoretic foundation for the two major pillars of modern physics, namely quantum theory (our framework for theories of the microscopic world) and general relativity (our best theory of the macroscopic world). One of the major stimuli for this program arises from the discoveries in black hole physics in the early 1970s, which have revealed a deep connection between the geometry of a black hole and its entropy (a measure of the information about the matter inside the black hole which we cannot access). Over the last few decades, this work has given rise to the striking notion that the outcomes of length and area measurements of space are inherently discrete, and that the information that can be stored in a region of space is intrinsically limited[5].

The informational viewpoint has generated at least two important strands of development which have impacted our understanding of quantum theory. First, the investigation of information processing within the context of quantum theory, which has given rise to the fields of quantum information and quantum computation. Second, the investigation of whether the quantum formalism itself can be characterized in terms of information-theoretic principles, and thus whether quantum theory can be understood as essentially a theory of information processing.

The classical theory of computation, formulated by Church and Turing in the 1930s, presupposes a classical conception of physical reality. All of our modern computers, however sophisticated, are based on this classical model of computation. The elementary unit of information is known as the *bit* (short for *b*inary dig*it*), which takes the value *zero* or *one* only. Abstractly, a computation is a deterministic map from a string of bits to another string of bits. More concretely, in an abstract computing device proposed by Turing, the device acts on a tape divided into square spaces, in each of which a symbol is printed. The device has a "head", which, at any time, is positioned opposite a particular space on the tape. At each time-step, the head either reads the symbol printed on the space, or erases and writes a symbol to the space; and the device moves the tape one unit to the left or right or leaves it in the same place. The device also has a discrete internal state, which is modified in a deterministic way at each time-step. The essential idealizations of classical physics are apparent here: there is no mention of quantum states and quantum measurements.

It is natural to wonder whether information processing fundamentally differs when embodied in quantum reality, and, in particular, whether new things are possible when the full richness of quantum reality is harnessed. It is also natural to wonder whether quantum theory imposes new constraints on how the information encoded in quantum systems can be manipulated.

The elementary unit of quantum information is not the bit but the *qubit*. Information is encoded in the *state* of the qubit, which can be visualized by a point on the surface of a sphere. According to quantum theory, a measurement performed on the qubit will yield only one of *two* possible outcomes. Thus, there is a strange disparity between the state of the qubit (which can take on a two-fold continuum of possible values), and the result of a measurement (which can take only two possible values). In particular, it follows from quantum theory that the state of a qubit cannot simply be "read" as can that of a classical bit.

However, one might imagine that there exists a process which is capable of *copying* the state of a qubit without actually reading it. One of the most striking early findings of quantum information in the early 1980's is that, according to quantum theory, this is impossible [8]. That is, unlike a classical bit, there is no general device which, when fed a qubit in an unknown state, can output two qubits in the same unknown state. But quantum information is not only about restrictions. Over the last thirty years, it has been found that, when information is embodied in quantum systems, one can carry out information processing tasks which are impossible or very difficult to carry out using classical information processing. For example, one can in principle build a *quantum computer* which can quickly solve certain problems (such as the problem of factoring large numbers) which are of great practical importance and yet are intractable on classical computers. One can also carry out tasks using quantum information processing that are classically *impossible.* For example, using quantum theory in what is known as *quantum cryptography,* two parties can communicate a message encoded in qubits, and yet to be able to detect whether there is an eavesdropper listening in on their communication. This sensitivity to eavesdropping relies essentially on the fact that, in quantum physics, unlike classical physics, a measurement (the eavesdropping) is an *active* process that, in general, affects what is being measured. Remarkably, most of the innovations allowed by information processing depend crucially upon the use of *entanglement.* For example, in quantum cryptography, it is essential that the two parties involved share entangled particles. For this reason, in quantum information processing, entanglement has come to be viewed as an indispensable resource. At a foundational level, this has stimulated the study of entanglement in its own right, and has persuaded many physicists that entanglement is a very real facet of physical reality.

---

[5] *Editor's note*: Within the context of, for example, loop quantum gravity, discussed elsewhere in the book.

RECONSTRUCTION OF QUANTUM THEORY

As mentioned earlier, the formalism of quantum theory was obtained in a rather *ad hoc* manner, as a result of which it possesses many mathematical features whose physical origin is unclear[6]. Presumably, there are *physical* reasons for these features, but what are they? Until the 1980s, most progress to understand the nature of quantum reality has been made by taking most or all of the quantum formalism as a given, and then attempting to draw physical implications from it in various ways. The early interpretation of de Broglie, Bell 's theorem of non-locality, and the discoveries of quantum information processing are all examples of this methodology. As a result of taking most of the formalism as a given, however, these approaches are intrinsically unable to account for the physical origin of the quantum formalism itself.

Over recent years, there has been an increasing realization that such a methodology is also intrinsically limiting in other ways. For example, it is now established that quantum reality has numerous non-classical features, such as the probabilistic nature of measurements, non-locality, and so forth. But, if we are to understand the nature of quantum reality as a whole, we need to understand how these separate features fit together as part of a single, overarching picture of reality. However, as long as the quantum formalism is taken as a given, we cannot know which of these non-classical features are fundamental and which are secondary or derived, and neither can we be sure that there are no other important non-classical features which remain to be discovered that are, in some way, essential to our understanding of quantum reality.

One way forward is to take a few of the non-classical features of quantum theory that we think are fundamental to how nature works, and to try to *derive* – or reconstruct – the quantum formalism from these features in a systematic way. If successful, such a reconstruction would show that the non-classical features we started with are indeed sufficient by themselves to account for the full richness of quantum reality. Ideally, a reconstruction of the quantum formalism would distill the physical content of the quantum formalism into a set of clear, intuitively graspable statements expressed in natural (everyday) language, thereby rendering the entire content of the quantum formalism available to the kind of conceptual analysis that is necessary to develop an underpinning conception of reality. Thus, reconstruction is a powerful methodology that can be used to *test* any candidate understanding of quantum reality, and, by focussing our attention on just a small number of the non-classical features of quantum theory expressed in natural language, facilitate the building up of a conception of physical reality capable of fully underpinning it.

INFORMATION-BASED RECONSTRUCTION

The fundamental role played by the concept of information in a discussion of the quantum formalism, together with the insights into the nature of quantum reality gained through an exploration of quantum information processing, naturally gives rise to the question of whether quantum theory can be reconstructed on the basis of information-theoretic principles. One of the earliest investigations in this direction was carried out by Wootters [9], who showed that a fundamental prediction (Malus ' law) of quantum theory could be derived by assuming that quantum reality is, in some precise sense, as good as it could possibly be for sending information via quantum systems from one party to another. Another early investigation [10] centers on the remarkable fact that, although quantum reality is non-local in the sense shown by Bell, quantum theory also implies that this non-locality cannot actually be used for instantaneous signaling (which would be in violation of special relativity). As physicists sometimes say, there is a peaceful coexistence between the properties of non-locality and no-signalling. Although this investigation concluded that this peaceful co-existence was not by itself sufficient to account for the structure of quantum theory, it was an important inspiration for later work.

Over the last decade, a number of detailed reconstructions have been presented, which proceed from diverse starting points. To illustrate the nature of a reconstruction in more detail, I will mention recent work in which I have shown how it is indeed possible to reconstruct the core of the quantum formalism (namely Feynman 's rules of quantum theory) [11]. By making use of elementary symmetry and consistency conditions, I have shown that it is possible to derive the complex nature of the quantum formalism directly from the probabilistic nature of measurement outcomes

[6] For example, the states of quantum systems are most naturally described using complex numbers, but the physical reason for this has not yet been identified.

and complementarity. This reconstruction strongly suggests that the quantum formalism can indeed be regarded as a way to process information consistent with particular informational constraints. This derivation takes place without making reference to more than one physical system, thus showing that features such as non-locality and no-signaling are *not,* in fact, essential to an understanding of the structure of the quantum formalism. That is, from the point of view of the derivation, features such as non-locality and no-signalling are *not* fundamental, but secondary. Moreover, the derivation makes no use of spatial locality, which suggests that the quantum formalism is *in some sense more fundamental than* space.

IMPLICATIONS FOR QUANTUM GRAVITY AND OUR CONCEPTION OF PHYSICAL REALITY

In the existing quantum theories of particles and fields, spacetime is treated classically, while the particles and fields themselves are described using the quantum formalism. Thus, these theories are curious hybrids where particles interact via influences propagating through space, but can also interact non-locally via entanglement-mediated influences which *bypass* space. Is there some more fundamental way to unify local and non-local interactions? As discussed in the preceding section, the quantum formalism can be derived without reference to space, which strongly suggests that the quantum formalism itself is inherently non-spatial. That is, quantum theory is neither non-local nor local; it is *more fundamental* than either notion. Nevertheless, the reconstruction also shows that entanglement *is* a basic feature of this formalism. This suggests that one should not regard locality (and thus space) and non-locality as primitive givens, but rather approximate features of the world, which jointly emerge from some more basic, pre-spatial, reality in which entanglement plays a central role. If valid, this view suggests that, if one wishes to develop a theory of quantum gravity, one should not start with space as a given and try to describe it directly using the quantum formalism, but rather to try to find some more primitive pre-spatial structure, from which space will emerge as an approximation, perhaps at large-length and low-energy scales.

If, as is suggested by the reconstruction, we accept indeterminacy and complementarity as a given, what conception of physical reality do they suggest? What are they telling us about how Nature works? A reconstruction of quantum theory, as the one referred to above, is an important stepping-stone to a new conception of nature in that it focuses our attention on just a small number of non-classical features. But it is still only a stepping-stone. How does one proceed beyond this point?

To start with, we need to understand indeterminacy and complementarity. For example, it is one thing to accept that measurements are indeterministic as an operational principle (that is, as a summary of what we find in our experiments), but quite another to accept at a philosophical level that things happen without any cause whatsoever. How are we to get a philosophical handle on such an idea? In the case of complementarity, why is it that a measurement can access *one-half* of the degrees of freedom of the state of a system, and not some other fraction, for instance? One of the important jobs of a conception of reality is to provide a unified understanding of the separate assumptions that underlie a reconstruction. In the above reconstruction, indeterminacy and complementarity have been postulated for essentially different reasons. So, in this case, we would like to find an overarching understanding of why Nature has *both* these features, and not just one or the other.

We must also face the measurement problem. In the above reconstruction (and also in almost all recent reconstructive attempts), the notion of measurement is taken as a given. Yet, as previously discussed, we do not know what kind of physical process constitutes a measurement. My own view is that the indeterministic, irreversible physical process we commonly refer to as "measurement" occurs constantly at the microscopic level of matter, and thus is an objective feature of reality's *unfolding* rather than (as one might suppose from the language of quantum theory) something that depends upon observers performing what we ordinarily think of as measurements. Accordingly, the challenge is to understand the conditions under which the dynamics of a system are dominated by deterministic dynamics (mathematically referred to as unitary dynamics), and when and why measurement dynamics (indeterministic, irreversible dynamics) occur. Currently, all of these issues remain open, and are the focus of current research.

CONCLUSION

As described in the Introduction, quantum theory poses deep challenges to the classical view of physical reality, and, by extension, to the mechanical world-view. At present, there is no consensus on how to meet those challenges, no consensus on how to revise the classical view of physical reality or the mechanical world-view which it spawned. Nevertheless, as I hope I have conveyed in the preceding pages, many vital insights into the nature of quantum reality have been obtained, chief amongst them the nature of entanglement and non-locality, and the role of information in the quantum realm.

Methodologically, the recent upsurge of interest in the reconstruction of quantum theory has the potential for unifying the insights that have been obtained thus far, and lead to the discovery of the key physical ideas that underlie the quantum formalism. Such a distillation of the physical content of the quantum formalism into a small number of simple physical ideas has the potential to provide the basis for the development of a coherent conception of physical reality that takes account of the full physical content of the quantum formalism, which is probably vital for the development of a more comprehensive theory of physics.

REFERENCES

[1] N. Bohr, "Causality and complementarity". Philosophy of Science, 4, pp289–298, 1937.

[2] A. Pais, *Niels Bohr's Times*, OUP, 1991.

[3] W. Heisenberg, *Physics and Beyond*, World Perspectives, 1971.

[4] W. Pauli, *Writings on Physics and Philosophy*, Springer, 1994.

[5] J. Bell, "On the Einstein Podolsky Rosen paradox", *Physics*, 1: pp195–200, 1964.

[6] J. Wheeler, "It from bit", *Proceedings of the 3rd International Symposium on the Foundations of Quantum Mechanics*, Tokyo, 1989.

[7] J. Wheeler, *Information, physics, quantum: The search for links*, *Complexity, Entropy, and the Physics of Information*,. Addison-Wesley, 1990.

[8] W. Zurek and W. Wootters, "A single quantum cannot be cloned", *Nature*, 299, pp802–803, 1982.

[9] W. Wootters, *The acquisition of information from quantum Measurements*, University of Texas at Austin, 1980.

[10] S. Popescu and D. Rohrlich, *Causality and nonlocality as axioms for quantum mechanics, Causality and Locality in Modern Physics and Astronomy: Open Questions and Possible Solutions*, 1997.

[11] P. Goyal, K Knuth, and J. Skilling, "Origin of complex quantum amplitudes and feynman's rules", Phys. Rev. A81, p022109, 2010.

INDEX

A

B

C

D

E

F

G

H

I

K

L

M

T

U

W

Y

Z

Made in United States
North Haven, CT
04 November 2026

21730007R00073